Prepared in cooperation with Maine Department of Transportation

Relations Among Water Levels, Specific Conductance, and Depths of Bedrock Fractures in Four Road-Salt-Contaminated Wells in Maine, 2007–9

Scientific Investigations Report 2012–5205

U.S. Department of the Interior
U.S. Geological Survey

Cover. Photograph shows South Sullivan project well.

Relations Among Water Levels, Specific Conductance, and Depths of Bedrock Fractures in Four Road-Salt-Contaminated Wells in Maine, 2007–9

By Charles W. Schalk and Nicholas W. Stasulis

Prepared in cooperation with Maine Department of Transportation

Scientific Investigations Report 2012–5205

U.S. Department of the Interior
U.S. Geological Survey

U.S. Department of the Interior
KEN SALAZAR, Secretary

U.S. Geological Survey
Marcia K. McNutt, Director

U.S. Geological Survey, Reston, Virginia: 2012

For more information on the USGS—the Federal source for science about the Earth, its natural and living resources, natural hazards, and the environment, visit http://www.usgs.gov or call 1–888–ASK–USGS.

For an overview of USGS information products, including maps, imagery, and publications, visit http://www.usgs.gov/pubprod

To order this and other USGS information products, visit http://store.usgs.gov

Suggested citation:
Schalk, C.W., and Stasulis, N.W., 2012, Relations among water levels, specific conductance, and depths of bedrock fractures in four road-salt-contaminated wells in Maine, 2007–9: U.S. Geological Survey Scientific Investigations Report 2012–5205, 47 p., at http://pubs.usgs.gov/sir/2012/5205.

Acknowledgments

The authors thank the private land owners who gave us access to their wells and welcomed the installation of instrument shelters in their front yards. Thanks to the Well Claims Unit at Maine Department of Transportation for field and data support during this project. Thanks to Carole Johnson of the USGS for her assistance in collecting and analyzing geophysical data.

THIS PAGE INTENTIONALLY LEFT BLANK

Contents

Figures

Table

Conversion Factors and Datums

Inch/Pound to SI

Multiply	By	To obtain
Length		
inch (in.)	25.4	millimeter (mm)
foot (ft)	0.3048	meter (m)
mile (mi)	1.609	kilometer (km)
Volume		
gallon (gal)	3,785	milliliter (mL)
gallon (gal)	0.003785	cubic meter (m^3)
gallon (gal)	3.785	cubic decimeter (dm^3)
Flow rate		
foot per minute (ft/min)	0.3048	meter per minute (m/min)
gallon per minute (gal/min)	0.06309	liter per second (L/s)

Temperature in degrees Celsius (°C) may be converted to degrees Fahrenheit (°F) as follows:

°F=(1.8×°C)+32

Temperature in degrees Fahrenheit (°F) may be converted to degrees Celsius (°C) as follows:

°C=(°F-32)/1.8

Vertical coordinate information is referenced to the North American Vertical Datum of 1988 (NAVD88).

Horizontal coordinate information is referenced to the North American Datum of 1983 (NAD 83).

Altitude, as used in this report, refers to distance above the vertical datum.

Specific conductance is given in microsiemens per centimeter at 25 degrees Celsius (µS/cm at 25°C).

Concentrations of chemical constituents in water are given either in milligrams per liter (mg/L) or micrograms per liter (µg/L).

THIS PAGE INTENTIONALLY LEFT BLANK

Relations Among Water Levels, Specific Conductance, and Depths of Bedrock Fractures in Four Road-Salt-Contaminated Wells in Maine, 2007–9

By Charles W. Schalk and Nicholas W. Stasulis

Abstract

Data on groundwater-level, specific conductance (a surrogate for chloride), and temperature were collected continuously from 2007 through 2009 at four bedrock wells known to be affected by road salts in an effort to determine the effects of road salting and fractures in bedrock that intersect the well at a depth below the casing on the presence of chloride in groundwater. Dissolved-oxygen data collected periodically also were used to make inferences about the interaction of fractures and groundwater flow. Borehole geophysical tools were used to determine the depths of fractures in each well that were actively contributing flow to the well, under both static and pumped conditions; sample- and measurement-depths were selected to correspond to the depths of these active fractures. Samples of water from the wells, collected at depths corresponding to active bedrock fractures, were analyzed for chloride concentration and specific conductance; from these analyses, a linear relation between chloride concentration and specific conductance was established, and continuous and periodic measurements of specific conductance were assumed to represent chloride concentration of the well water at the depth of measurement.

To varying degrees, specific conductance increased in at least two of the wells during winter and spring thaws; the shallowest well, which also was closest to the road receiving salt treatment during the winter, exhibited the largest changes in specific conductance during thaws. Recharge events during summer months, long after application of road salt had ceased for the year, also produced increases in specific conductance in some of the wells, indicating that chloride which had accumulated or sequestered in the overburden was transported to the wells throughout the year. Geophysical data and periodic profiles of water quality along the length of each well's borehole indicated that the greatest changes in water quality were associated with active fractures; in one case, high concentration of dissolved oxygen at the bottom of the well indicated the presence of a highly transmissive fracture that was in good connection with a surficial feature (stream or atmosphere). Data indicated that fractures have a substantial influence on the transport of chlorides to the subsurface; that elevated specific conductance occurred throughout the year, not just when road salts were applied; and that chloride contamination, as indicated by elevated specific conductance, may persist for years.

Introduction

In 1969 and also in 1987, the State of Maine legislature signed into law a process by which any landowner believing his or her water supply to be adversely affected by public construction, reconstruction, or maintenance of roads can apply to the political subdivision for redress (23 M.R.S.A. §652 (Maine Legislature, 2012, p. 60–61) and §3659 (Maine Legislature, 2012, p. 227–229)). Adverse effects can include those from road salting (University of Maine, School of Law, 2003; Maine Department of Transportation, 2004; State of Maine Judicial Branch, 2008), and redress can include replacing plumbing and (or) heating systems, replacing the water supply, repairing damage to the water supply, paying a designated sum of money, and (or) purchasing the real estate served by the water supply.

Such remedial efforts have been expensive. In complying with the law, the Well Claims Unit at Maine Department of Transportation (MaineDOT) has handled almost 450 well claims since 1990, of which 130 have been related to road-salt contamination.[1] Claims for road-salt contamination have

---

[1] The secondary maximum contamination level for chloride is 250 milligrams per liter (U.S. Environmental Protection agency, 2012). Background concentrations of chloride in groundwater in the crystalline bedrock areas of New Hampshire and Maine are in the range of 20 to 30 mg/L (New Hampshire Department of Environmental Services, 2010; Maine Interagency Report, 2001).

grown from about 1 in 1990 to as many as 23 in 2006 (though not all of these claims were valid). Costs to the State to remediate a valid claim usually range from $10,000 to $35,000, and this amount can be much greater if construction (such as laying municipal water-supply lines) or the purchase of real estate is needed (Joshua Katz, Maine Department of Transportation, written commun., 2009).

The process of evaluating well claims includes an assessment of groundwater quality, particularly the concentrations of sodium and chloride. MaineDOT has kept long-term water-quality records from several wells, including some that were part of the well-claim process and others that are on publicly owned land and were suitable for monitoring. As of 2007, data collected by MaineDOT indicated that concentrations of chloride in groundwater spiked annually in the autumn, before road salts were applied for the winter and as groundwater levels were recovering from summer declines. This observation led MaineDOT to question whether chloride concentrations were increasing with time in groundwater in fractured bedrock in Maine as an outcome of general road salting practice, and the extent to which fractures were contributing to the apparent increases in chloride concentrations in the autumn.

The data collected by MaineDOT over the years had two shortcomings: first, the samples were discrete in time, being collected quarterly or less frequently; and second, each sample was mixed over the length of the borehole from which it was collected. To address these shortcomings, MaineDOT and the U.S. Geological Survey (USGS) began an investigation in 2007 into the nature and timing of the processes whereby road-salt constituents, particularly chloride, move into groundwater in fractured bedrock. By use of continuous monitoring and focusing on measurements as a function of depth, this investigation sought to determine whether the changes in chloride concentrations are related to the presence of fractures and hydrologic stresses, such as recharge and pumping.

Purpose and Scope

The purpose of this report is to enable concerned parties to understand the nature of the road-salt problem in parts of Maine. The report describes the occurrence of road-salt constituents in groundwater in selected bedrock wells in Maine and presents an evaluation of those occurrences with respect to fractures in the bedrock and timing of recharge events. This report does not address the theoretical aspects of flow in fractures but rather uses the body of literature and data observed from 2007 through 2009 to make general inferences about the roles of fractures in the transport of road-salt constituents.

Although this report presents parts of the data collected during the project, primarily those that are most important to the analyses described herein, complete data sets (discrete and continuous water levels, water temperatures, and specific conductance) are available online through the USGS Web site (http://waterdata.usgs.gov/me/nwis/gw or http://wdr.water.usgs.gov/) for water years 2008 (October 1, 2007 through September 30, 2008) and 2009 (October 1, 2008 through September 30, 2009) or by request from the Maine Water Science Center.

During the 18 months to 2 years in which data were collected, the hydrologic events that occurred may not have been typical. Because of the highly variable nature of hydrologic events, local and regional fracturing in bedrock, and differences in bedrock composition throughout Maine, the results described in this report may not be transferable to locations other than those studied.

Background

Salt, most commonly in the form of sodium chloride (NaCl), began to be used in the United States as a deicing agent about 1940. The use of salt for deicing increased through 2008, when it reached a peak of about 25 million tons (fig. 1), or about two-thirds of all salt used for all purposes in the

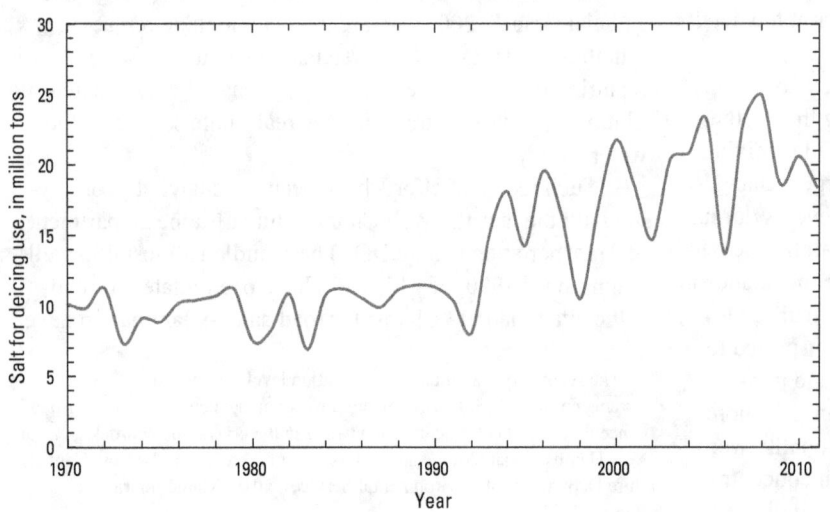

Figure 1.   Salt use for deicing roads in the United States since 1970. Data are from Kostick (1994, 1996–2012) and U.S. Geological Survey (2010).

United States. Although annual use for deicing is variable, the amounts used in 2005 and 2008 are more than twice that used in 1990. The data in figure 1 are not adjusted for increases in miles of road or severity of winters.

The State of Maine's use of salt for deicing (fig. 2) also has doubled since 1990, increasing from about 70,000 metric tons per year (t/yr) to about 140,000 t/yr. The data in figure 2 also do not take into account the severity of winters. Corresponding to the increase in salt use, however, has been a 98 percent reduction in the volume of sand applied to State highways since 1990. These trends reflect MaineDOT's policy toward anti-icing (proactive maintenance which does not use sand) rather than deicing, which has resulted in not only better service for the public (clear roads in short times) but also significant cost savings (Joshua Katz, Maine Department of Transportation, written commun., 2009). MaineDOT maintains about 18 percent of Maine's roads, so total salt use in the State is larger than the amounts shown in figure 2, perhaps about 490,000 t/yr (Rubin and others, 2010).

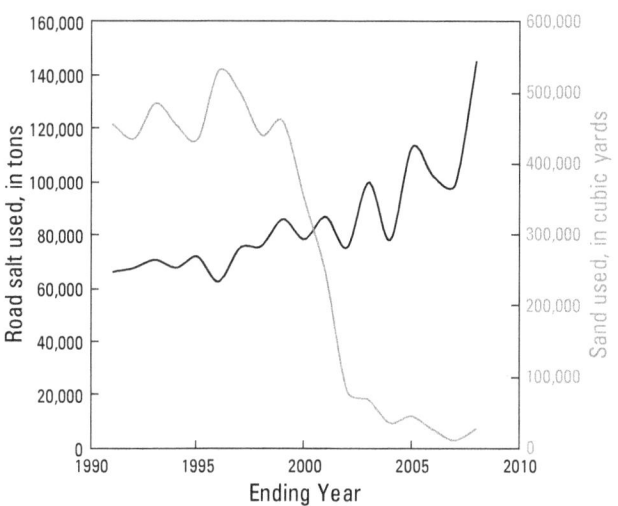

Figure 2. Salt and sand use for deicing roads by Maine Department of Transportation, 1990–2008. Data are from B. Burne, Maine Department of Transportation, written commun., 2009.

In the 1980s, MaineDOT covered or otherwise engineered their salt and sand piles to minimize the extent to which improper storage contributes to local groundwater quality. Olson and others (1985) indicated that improper storage had contributed to chloride concentrations as much as 2,900 milligrams per liter (mg/L) in one bedrock well. After engineered solutions were put in place, concentrations were lowered to 5 to 20 mg/L, much nearer to typical background levels.

Previous Investigations

Concerns about road salt as a potential contaminant in drinking water date to the 1950s, when it was discovered that salt was contaminating drinking-water supplies because of improper storage and, in some cases, highway runoff (Transportation Research Board, 1991). Research has shown that road-salt constituents, including anticaking agents such as iron cyanide (Paschka and others, 1999), continue to be present in streams and wetlands long after the winter season has passed (Demers and Sage, 1990), to the detriment of natural ecosystems and safe consumption of water (Andrews, 1996; Williams and others, 1999; Forman and Deblinger, 2000; Richburg and others, 2001; Blasius and Merritt, 2002; Godwin and others, 2003; Corsi and others, 2010). In some locations including the northeastern United States, those constituents are increasing in concentration in surface waters (Rosenberry and others, 1999; Nimiroski and Waldron, 2002; Jackson and Jobbagy, 2005; Kaushal and others, 2005). In their review of the effects of chemical deicers on the environment, Ramakrishna and Viraraghavan (2005) summarized the effects of sodium chloride on surface water: increasing density-driven gradients, increasing chloride concentrations, increasing salt-induced stratification, and stimulating algal growth.

The connection between application of deicing chemicals and the quality of groundwater is not difficult to deduce. Although streams can carry high loads of road salts away from an area, especially during runoff events (Ostendorf and others, 2009; Corsi and others, 2010; Harte and Trowbridge, 2010), some of the applied salts find their way into underlying aquifers during the recharge process (Blomkvist and Johanssen, 1999). Studies have shown that although the presence of road salts in groundwater may be masked by time, dilution, and spatial distribution, concentrations in groundwater tend to be relatively stable, enduring perhaps for decades after salting ceases (Williams and others, 1999; Foos, 2003; Bester and others, 2006; Ostendorf and others, 2006; Harte and Trowbridge, 2010). Some research has shown that splashing, plowing, and air currents transport 20 to 63 percent of road salts by weight to land surfaces as much as 120 feet (ft) from the road (Blomkvist and Johanssen, 1999), from which the salts may be carried into surface-water drainage or infiltrate the subsurface. The presence of road salts in groundwater has been studied most frequently in unconsolidated deposits near roads (Granato and others, 1995; Aichele, 2004; Kunze and Sroka, 2004; Andrews and others, 2005). Gradient-driven recharge from the surface and horizontal groundwater flow tends to dilute the salts, thereby seemingly alleviating any acute problems (Methuel, 2007). However, the effects also are noticeable in bedrock environments, where fractures and bedding planes greatly influence flow volumes and directions (Foos, 2003). Although chloride may be present naturally in bedrock minerals and aquifers (Aichele, 2004), its presence has been linked to road salting in urban and rural areas; chloride tends to be mobile in the environment, whereas sodium tends to adhere to soil particles. Groundwater contamination by road salt appears to be present primarily near road surfaces and in direct proportion to the application rate (Foos, 2003), even if the flow paths are unknown.

Because groundwater flow rates are much lower than those of streams, the problem of salt-contaminated groundwater often is chronic rather than acute. Salts can be present in the aquifers for years or decades, often at concentrations greater than are found in samples from monitoring wells. If the practice of deicing by use of road salts continues, concentrations in the aquifers can increase (Bester and others, 2006; Kelly, 2008). Generally, chloride-contaminated groundwater becomes diluted with distance from its source as it mixes with fresh recharge over time; however, dilution may not be an effective process in bedrock wells, depending on the groundwater flow path and the open interval of the well. In bedrock aquifers, where hydraulic conductivities usually are lower than those in unconsolidated materials, the residence time of contaminants may be decades or longer. Although, because of low hydraulic conductivities, it may take a long time for salt concentrations in groundwater to reach a level unfit for consumption, it also takes a long time for the contamination to be remediated (Rubin and others, 2010). Recognizing this problem, Bester and others (2006) recommend that "every effort should be made to reduce or eliminate [road] salting where possible," whereas Rubin and others (2010) state that "new policies are needed to encourage the use of chemicals and technologies that have fewer environmental effects than those of sand and salt."

Description of Study Areas and Study Wells

Most of Maine is underlain by crystalline bedrock of variable composition capped with shallow glacial materials (till, moraines, outwash) (Osberg and others, 1985; Thompson and Borns, 1985). Granite is the primary rock in about one-fourth of the State, and metamorphosed rock in the other three-fourths of the State. In general, the bedrock in Maine would be incapable of yielding sufficient water for domestic supply except for the presence of transmissive fractures in

the upper 500 ft. The presence of fractures in the bedrock is ubiquitous throughout the State, and any bedrock well yielding 2 or more gallons per minute (gal/min) almost certainly is intersected by one or more fractures (Caswell, 1979). Zones of highest-yielding bedrock wells in the State coincide with locations of faults and other geologic structures.

Four wells, each of which had to meet the following criteria, were chosen for the study (table 1). These criteria included:

- open-hole completion in bedrock;
- known to be affected by road salt;
- accessible by permission of the owner;
- along a State-maintained road.

Preliminary investigations by MaineDOT in response to landowner claims were used to determine the first two criteria. At the time a well claim is filed, MaineDOT samples the well for basic chemistry and salt-related constituents, including sodium, chloride, and bromide (Snow and others, 1990). Waters having a bromide signature are considered indicative of trapped seawater (natural brine contribution), whereas those lacking bromide, but still containing a salt signature, are considered indicative of anthropogenic sources that can include road salting, septic systems, or water softeners. Depth to bedrock, hydraulic gradient, and concentrations of nitrates also are considered when MaineDOT assesses the potential sources of contamination to wells.

The four wells chosen for this study were in the towns of Gray, West Gardiner, and Sullivan, Maine (fig. 3). Three of the four wells had been or were being used for domestic supply; the fourth was drilled for use as a domestic supply but had not been activated by the time the project ended. Drillers' logs were not available for any of the study wells, but geophysical logs were obtained for each well (appendix 1). Although construction data were available only for the south Sullivan well, geophysical data indicated that the casings of all wells extended below the bottom of the overburden. All wells were 6-inches in diameter, cased in steel, and open holed.

Table 1. Characteristics of wells used during the road-salt study in Maine, 2007–9.

[All coordinates are given in decimal degrees, North American Datum of 1983; --, not applicable]

Well identifier	Latitude, North	Longitude, West	Town	Depth, in feet below land surface	Casing height, in feet above land surface	Casing depth, in feet below land surface	Distance to nearest State road, in feet	Distance to Maine Turnpike, in feet
435103070191701	43.8508	-70.3214	Gray	327	1.0	75	40	380
441226069502201	44.2072	-69.8395	West Gardiner	347	1.5	25	160	1,200
443145068131801	44.5292	-68.2217	Sullivan	97.2	.5	15	25	--
443113068115101	44.5200	-68.1981	Sullivan	245	2.0	40	150	--

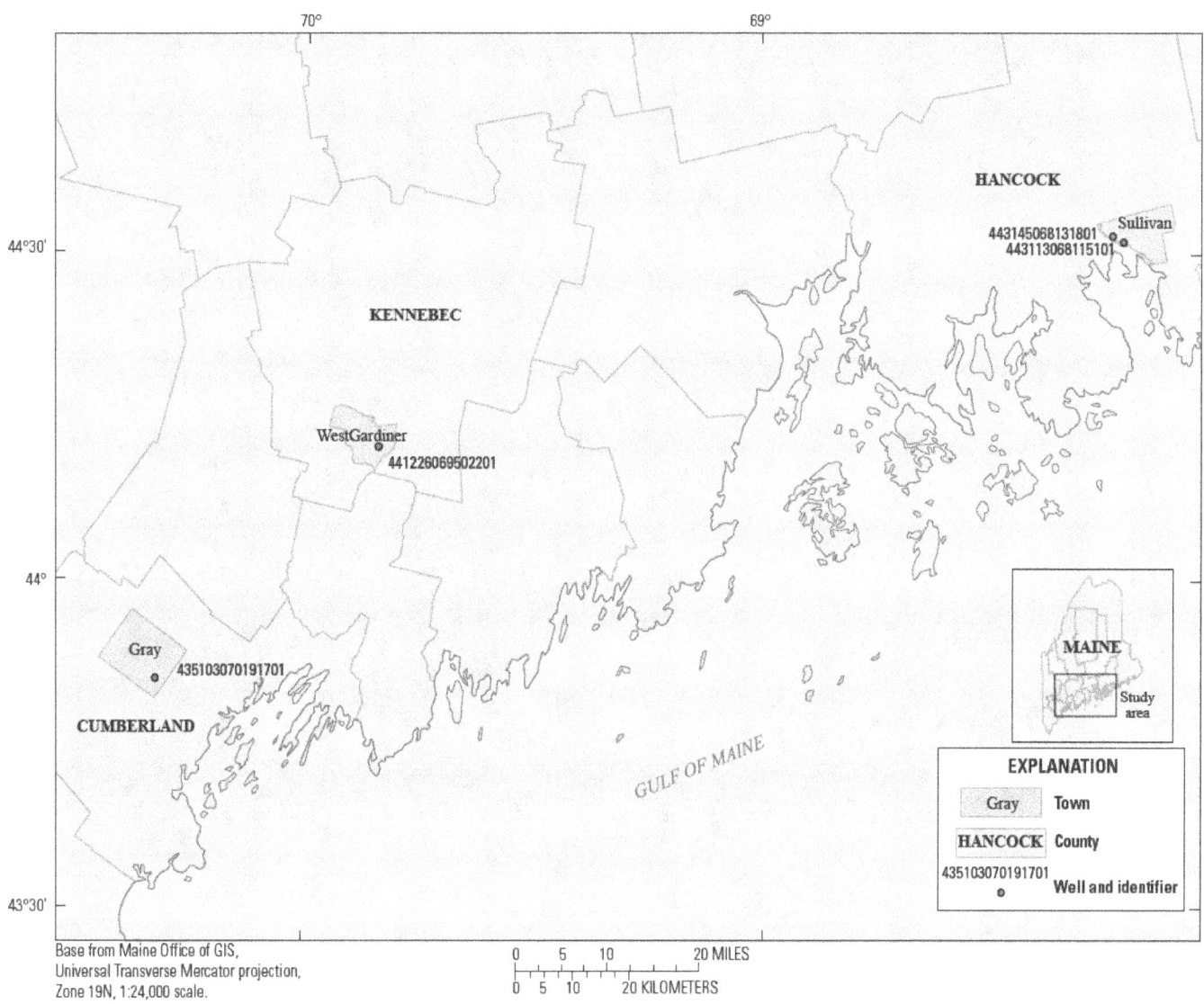

Figure 3. Locations of towns and wells used in the road-salt study, Maine, 2007–9.

Except for the south Sullivan well, which was grouted during construction, grouting (providing a surface seal to prevent surface-water runoff into the borehole) of the study wells was unknown.

The study area in Gray is underlain with carboniferous muscovite-biotite granite capped with glacial till, which generally is massive but can contain beds and lenses of stratified sediments (Osberg and others, 1985; Thompson and Borns, 1985). Geophysical logs show that the overburden is less than 75 ft thick (appendix 1). Groundwater-flow direction is assumed to be west to east, following the contours of the land surface (fig. 4). USGS well 435103070191701 is 380 ft downgradient from Interstate 95 and 40 ft from State Routes 26/100 to the west. The 327-ft-deep well formerly was used

for domestic supply, but because this and some other domestic wells along State Routes 26/100 south of Whitney Road were contaminated with road salts, MaineDOT and the Maine Turnpike Authority (MTA) developed infrastructure to bring municipally supplied water to landowners.

The study area in West Gardiner is underlain by the Silurian-age Waterville Formation, which consists of thinly bedded, gray to greenish-gray slate or pelitic schist with some wacke and calc-silicate rocks (Osberg, 1968; Osberg and others, 1985; U.S. Geological Survey, 2009b). Surficial sediments are fine-grained facies of glaciomarine origin, commonly clayey silts (Thompson and Borns, 1985). Groundwater-flow direction is assumed to be southeast to northwest, following contours of the land surface toward Cold Stream (fig. 5).

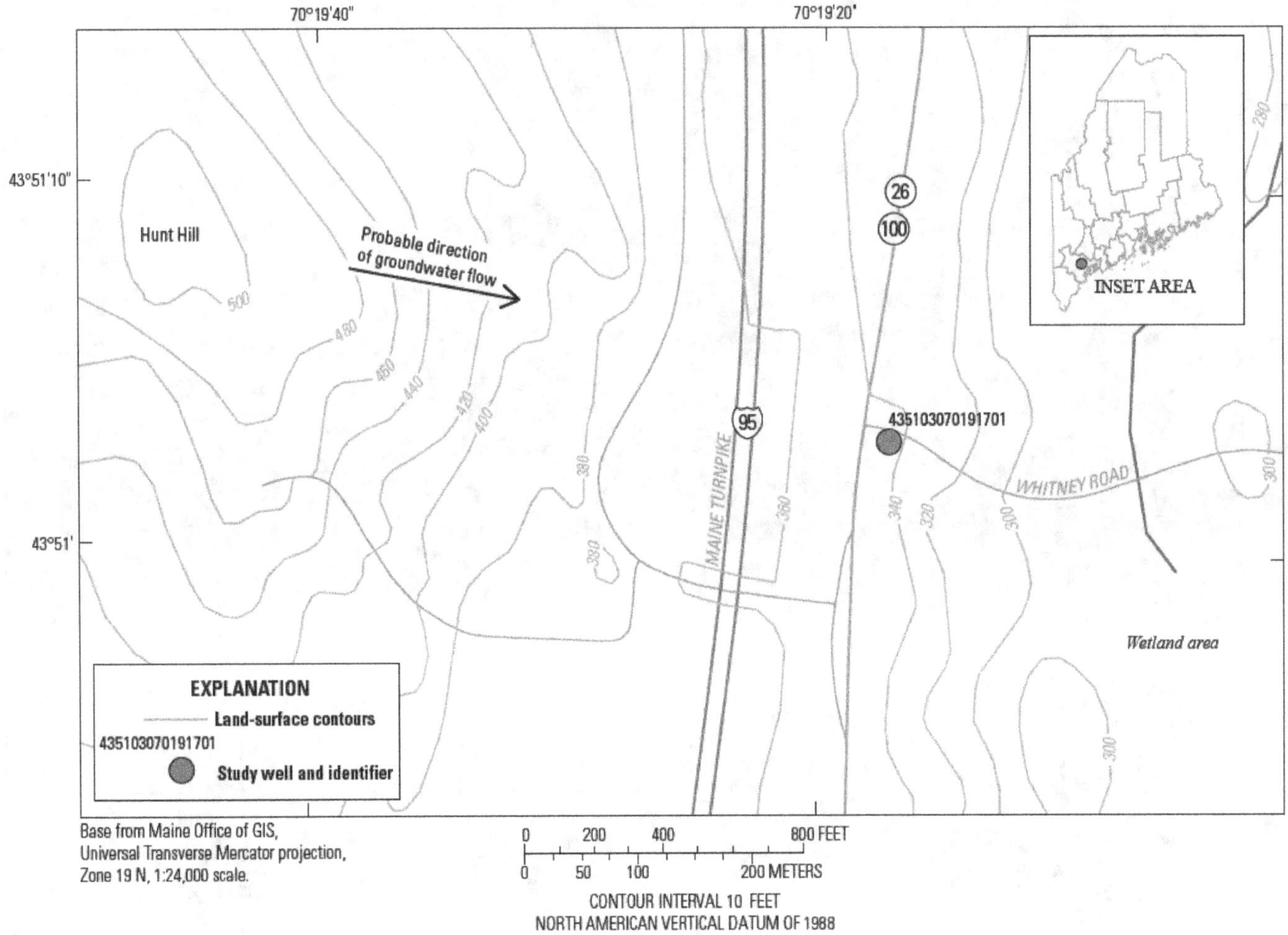

Figure 4.   Location of U.S. Geological Survey well 435103070191701 in Gray, Maine.

USGS well 441226069502201 is within 1,200 ft of a tollbooth station on Interstate 95 to the southeast and 160 ft of State Routes 9/126 to the north. The MTA has a maintenance yard at the intersection of Interstate 95 and State Routes 9/126 less than 0.5 mile (mi) east of the well. MaineDOT has a maintenance yard about 700 ft southwest of the well; the maintenance yard included an open salt pile from 1968 through 1988, when it was covered (E. Kluck and P. Coughlan, Maine Department of Transportation, written commun., 2009). The 347-ft-deep well formerly was used for domestic supply, but because this and some other wells nearby were contaminated with chloride, MaineDOT drilled a new supply well for the homeowners.

The study area in Sullivan is underlain with the quartz-granofels member of the Ellsworth Formation of Cambrian age (Osberg and others, 1985; Stewart, 1998). Surficial sediments are fine-grained facies of glaciomarine origin, commonly clayey silts (Thompson and Borns, 1985). Geophysical logs showed the overburden is less than 40 ft thick

(appendix 1). Groundwater-flow directions near the wells are uncertain, but probably follow contours of the land surface toward the bay (fig. 6). The 97-ft-deep North Sullivan well (USGS 443145068131801), which was still in use by the homeowner during the project, is within 25 ft of and at about the same elevation as U.S. Route 1 (maintained by MaineDOT) and within 530 ft of Taunton Bay. The South Sullivan well (USGS 443113068115101) is within 150 ft of and slightly uphill from U.S. Route 1 and within 370 ft of Taunton Bay. The 245-ft-deep South Sullivan well was drilled by MaineDOT to serve as a replacement water supply for a homeowner but was not activated for water supply until after the end of the project.

Climate

Average annual snowfall in Maine ranged from about 70 inches (in.) along the southern coast to 111 in. in the north during 1940–2002 (National Oceanic and Atmospheric

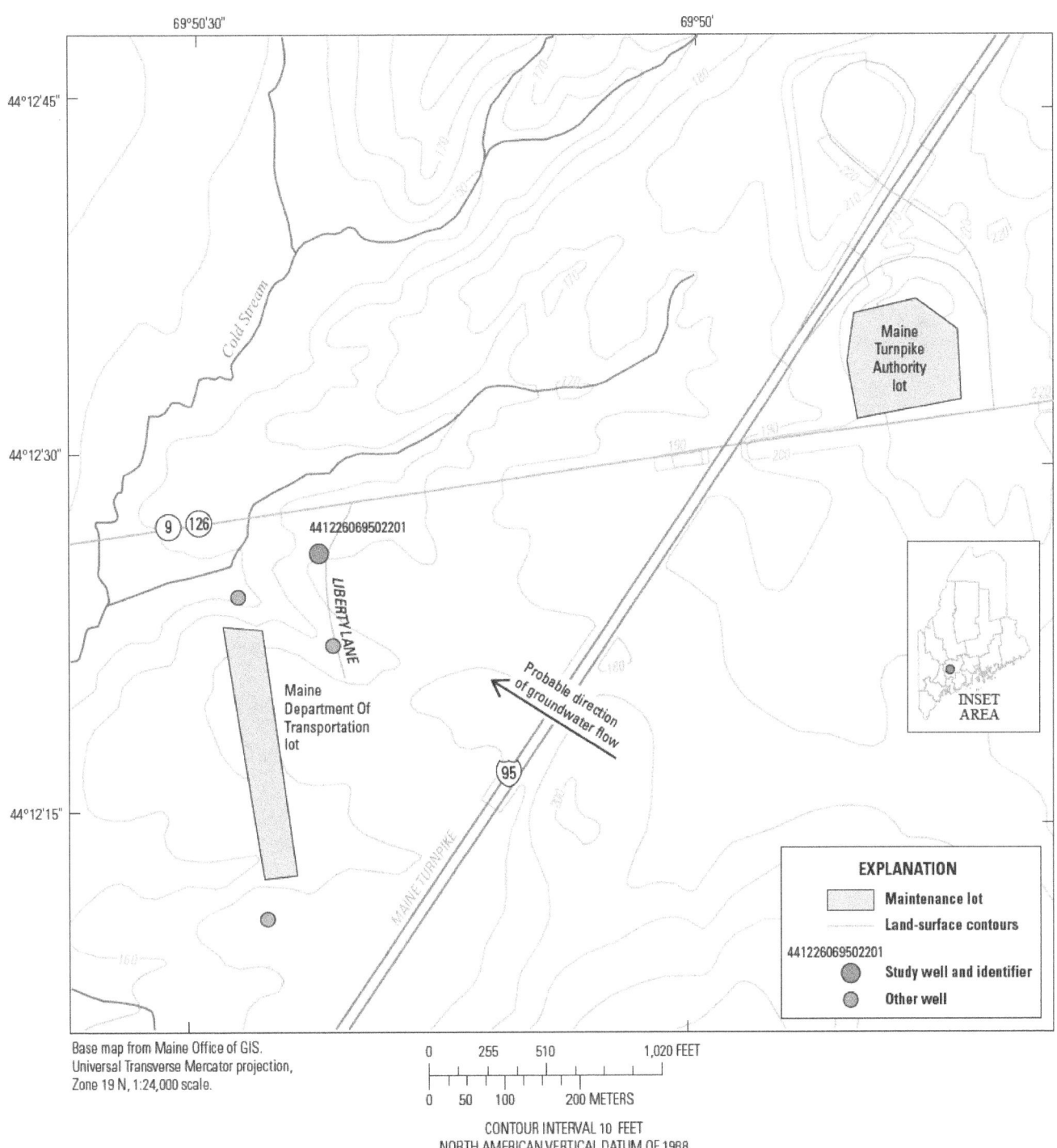

Figure 5.   Location of U.S. Geological Survey well 441226069502201 in West Gardiner, Maine.

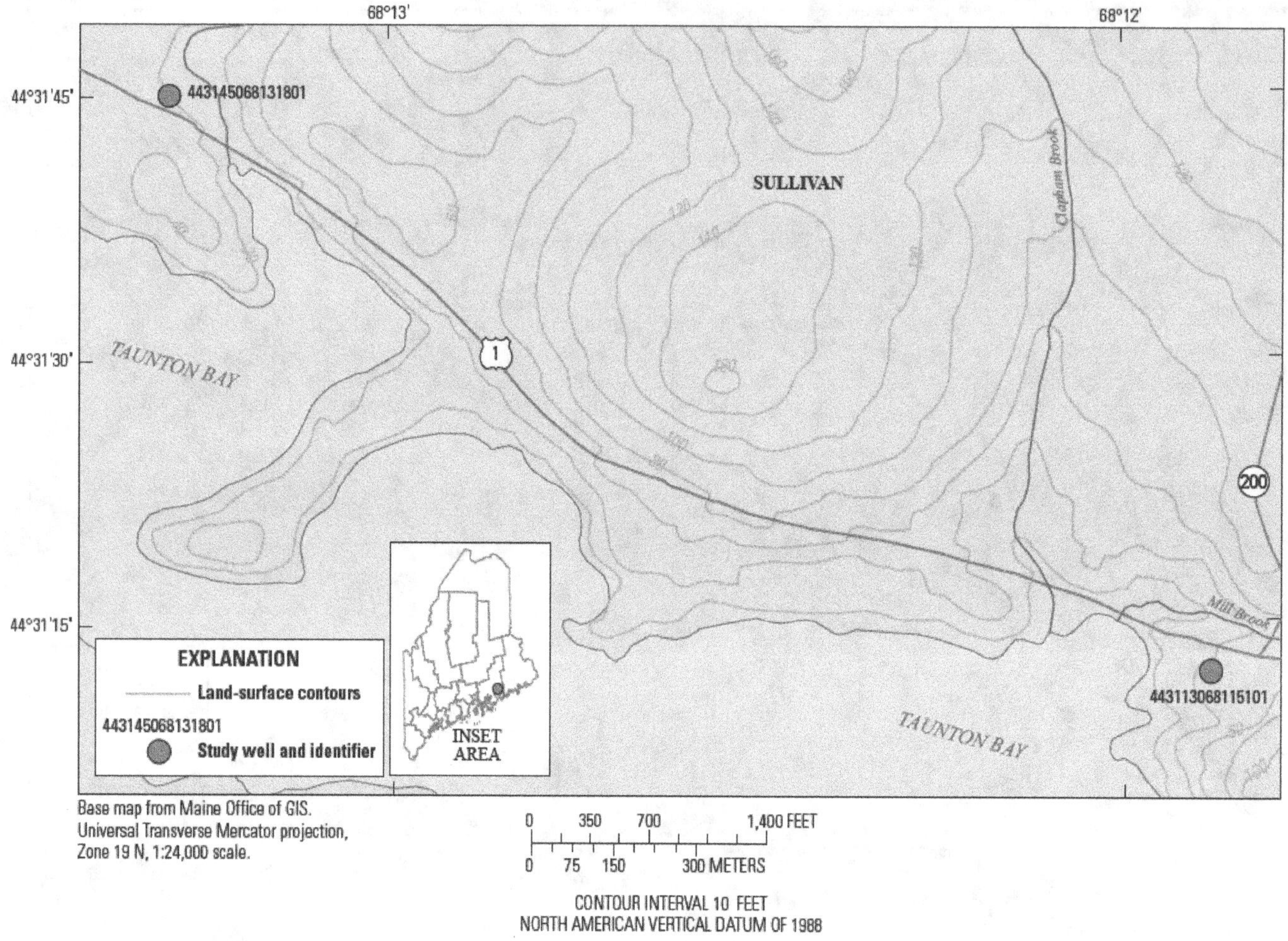

Figure 6.   Locations of U.S. Geological Survey wells 443145068131801 and 443113068115101 in Sullivan, Maine.

Administration, 2002) and 1895–1935 (Fobes, 1942). The snowiest years since 1997 at the National Oceanic and Atmospheric Administration (NOAA) observation station in Gardiner, Maine, were 2007 and 2009 (fig. 7). Average annual precipitation is about 45.8 in. in midcoastal Maine. Summer 2009 (June through August) was the wettest (22.31 in.) on record in Portland, Maine. Precipitation totals for 2005 (66.45 in.), 2006 (60.86 in.), 2008 (61.24 in.), and 2009 (58.61 in.) were among the eight highest ever recorded in Portland (National Weather Service, 2011).

Departures from normal monthly winter (December–April) temperatures during 1997–2009 at the NOAA observation station in Gardiner, Maine, are shown in figure 8. In general, the winter temperatures during 1997–2002, 2006, and 2008 were higher than normal, whereas the winter temperatures during 2003–5, 2007, and 2009 were lower than normal. The normal daily range in January temperatures in Gardiner is 4 to 28 degrees Fahrenheit (°F), whereas the normal daily range in April temperatures is 30 to 52 °F.

Methods of Data Collection

Standard methods were used to collect the water-level data (Cunningham and Schalk, 2011). All wells were instrumented with continuous water-level, temperature, and specific conductance monitors for periods of at least 15 months, including two spring snowmelt seasons. Data collected periodically included water-quality samples, water-quality profiles, and geophysical logs.

Continuous Water Levels and Water-Quality Data

The four wells were equipped with instruments to record and transmit continuous water levels, specific conductance, and water temperature. For the duration of the project, instruments for each well were housed in ventilated, waterproof, and locked aluminum shelters (fig. 9A). Instruments included a Design Analysis (DA) H522 data-collection platform (DCP)

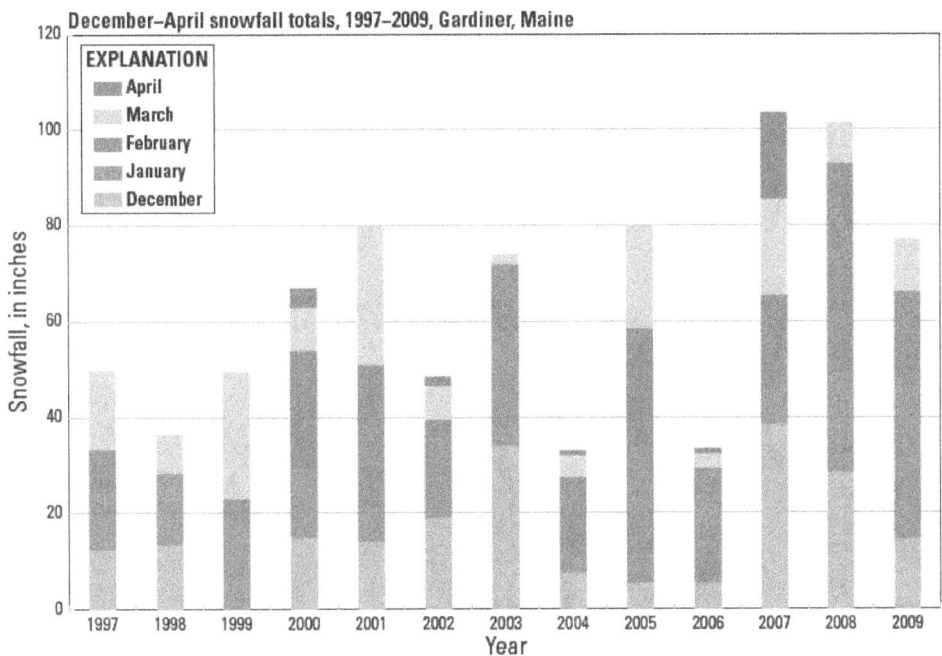

Figure 7.   December–April snowfall totals in Gardiner, Maine, 1997–2009.

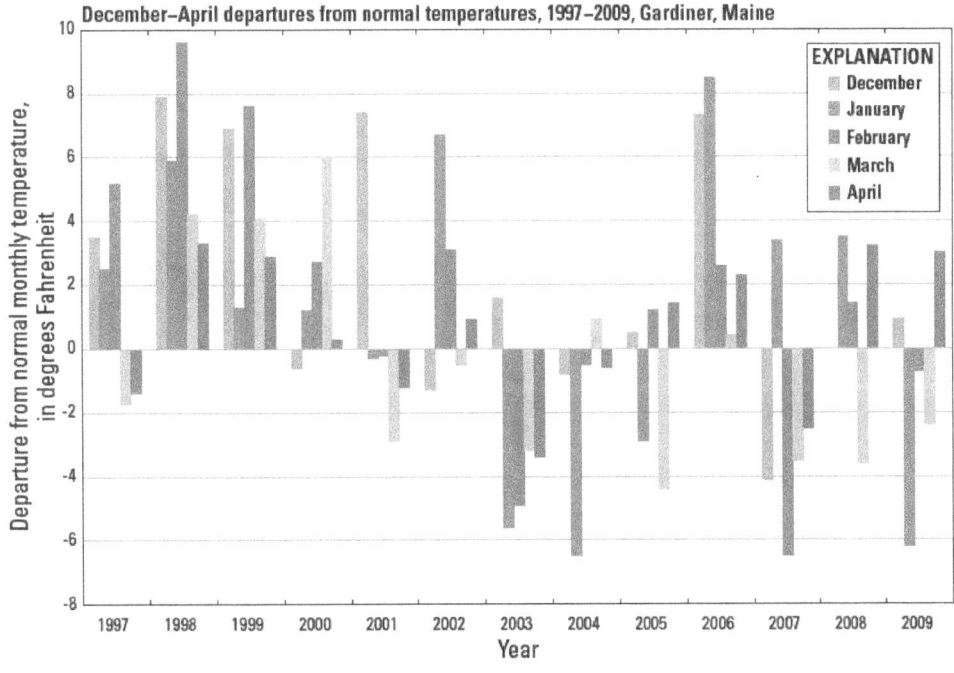

Figure 8.   December–April departures from average temperatures in Gardiner, Maine, 1997–2009.

A

B

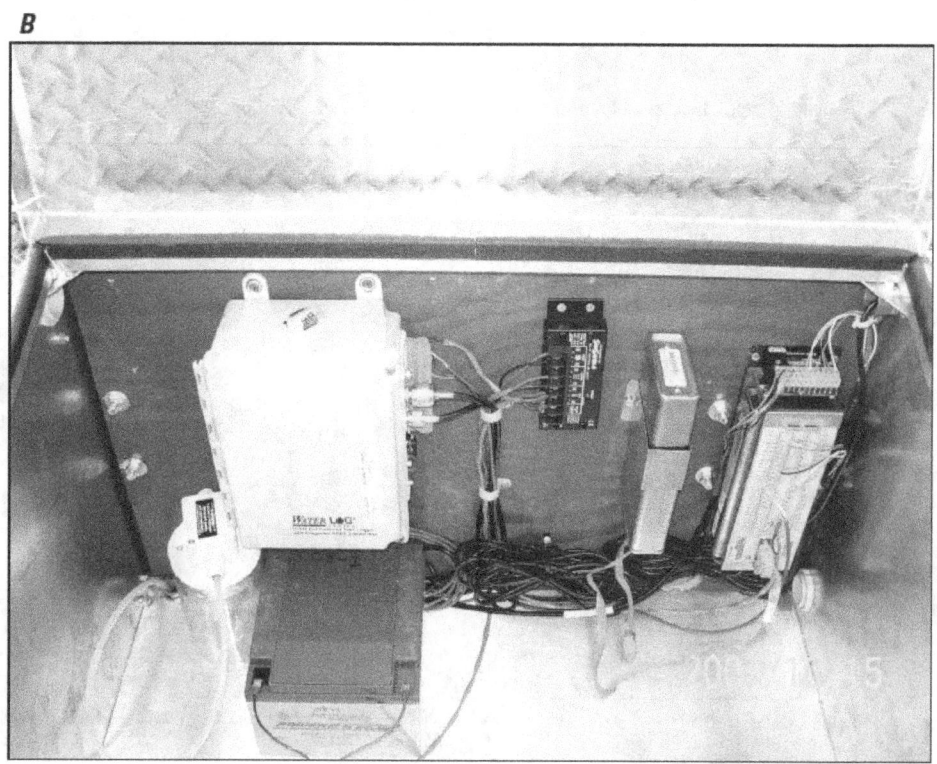

Figure 9.   *A,* Instrumentation shelter and *B,* arrangement of data-collection instruments at U.S. Geological Survey well 441226069502201 in West Gardiner, Maine.

attached to a DA H310 submersible pressure transducer and a Campbell Scientific (CS) CR10 DCP attached to a CS H547A temperature and conductance probe (fig. 9B). The DCP transmitted 15-minute water-level, water-temperature (from both the pressure transducer and the conductance probe), and specific-conductance data every hour over the geostationary satellite (GOES) system. The CS H547A temperature and conductance probe measures water temperature by use of a resistor that responds rapidly to changes in temperature but measures conductance only as water moves through the probe. If little to no vertical flow is present in the well, this instrument can indicate slow to no response to changes in conductance.

Specific conductance and temperature probes were checked for calibration against a range of calibration standards before deployment in the field, at least once during deployment, and when the probes were removed from the wells. All probes met manufacturer specifications during each calibration check.

Water levels were collected to 0.01 ft in all wells and were reported as depth below land surface datum (LSD). Temperature originally was collected to 0.1 degree Celsius (°C) in all wells, but after about December 7, 2007, it was collected to 0.01 °C. Specific conductance was recorded to three significant figures.

Gray

Installation of the monitoring instruments in the Gray well began on December 12, 2007, and concluded December 14, 2007. The pressure transducer was set at a depth of 42 ft below LSD and the conductance probe was set at a depth of 90 ft below LSD. Both instruments were replaced soon after installation was completed on December 14, 2007, because of issues with instrumentation catching on the pump in the well. The pump was removed the same day. Specific conductance data from December 2007 through January 2008 were considered unreliable and are not used in this report. All of the instruments were removed from the well and data collection ended on May 14, 2009.

West Gardiner

On August 30–31, 2007, a CS CR10 DCP was installed in a plastic shelter mounted over the well casing. The DCP was connected to a CS temperature and conductance probe at a depth of 44 ft below LSD. A DA DH21 pressure transducer also was installed in the well at a depth of 30 ft below LSD. This installation was to collect data temporarily until the real-time equipment was installed.

On October 15, 2007, the West Gardiner well was converted to a permanent installation. The pressure transducer was set at a depth of 42 ft below LSD and the conductance probe at a depth of 41 ft below LSD. On November 2, 2007, the pressure transducer was raised to a depth of 34 ft below LSD to prevent possible issues with the probes catching on

one another. The instruments were removed from this well and data collection ended on November 6, 2009.

North Sullivan

Instruments to measure water level and temperature were installed in the North Sullivan well at depths of 18 ft and to measure specific conductance and temperature at 80 ft on September 24, 2007. Because the data from winter 2008 showed high variability, possibly related to the effect of the daily use of water from the well, a second probe for specific conductance and temperature was installed in the well at a depth of 40 ft on June 4, 2008. The instruments were removed from this well and data collection ended on June 3, 2009.

South Sullivan

On January 10, 2008, the pressure transducer in the South Sullivan well was set at a depth of 45 ft below LSD and the conductance probe set at a depth of 42 ft below LSD. On June 4, 2008, the conductance probe was replaced with a new H547A probe set at a depth of 220 ft below LSD. These instruments were removed and data collection ended on June 3, 2009.

Water-Quality Samples and Profiling

Water-quality samples were collected to determine a relation between specific conductance, which was monitored continuously, and concentrations of chloride. Water samples were collected by use of diffusion samplers. The diffusion medium, from Spectrum Laboratories, Inc., was molecular porous cellulose membrane tubing with a 20.4-millimeter (mm) diameter and a molecular weight cutoff of 12,000 to 14,000 daltons.

In the USGS laboratory, samplers were filled with deionized water, sealed at both ends, and immersed in a deionized water bath for transport to the field. In the field, samplers were lowered to the appropriate depth and allowed to equilibrate under environmental conditions for at least a week. When a sampler was removed, the sample was poured immediately into a labeled 125 milliliter (mL) poly bottle and taken to the USGS laboratory. Specific conductance was measured in the laboratory by use of an Orion model 160 conductivity meter before the sample was sent to the USGS National Water-Quality Laboratory for analysis of chloride. The reporting limit for chloride concentration was 0.12 milligrams per liter (mg/L).

Vertical profiling of specific conductance, water temperature, dissolved-oxygen concentration, and pH was done at least once in each well. For this process, a rope of length greater than the depth of the well was attached to a calibrated, internally logging water-quality probe from Hydrolab or YSI. The rope was marked every 10 ft. While one hydrographer kept notes of the clock time and length of rope deployed, the other hydrographer lowered the probe through the well bore, stopping for a minute at each 10- or 20-ft increment to allow the probe to equilibrate and record. Probe calibrations were checked at the end of each day of profiling; in all cases, the probes remained within the calibration settings. Data from the probe were paired with corresponding depth data recorded in the field on the basis of clock time.

Geophysical Data

The USGS used conventional and advanced geophysical methods to characterize the wells and to provide information about the types of rock at each well location, orientation of bedding planes, degree of fracturing, and vertical flow directions of water in each well. Conventional logs collected in each well included caliper, natural-gamma, specific conductance (fluid-resistivity), and fluid-temperature logs. Advanced methods included optical and acoustic televiewer logging and heat-pulse flowmeter logging. Descriptions of these methods are from Johnson and others (2002).

Caliper logging generates a continuous profile of borehole diameter with depth. The caliper tool is pulled up the borehole, allowing three spring-loaded arms to open as they pass borehole enlargements. The enlargements generally are related to fractures but can also be caused by changes in lithology or borehole construction.

Natural-gamma logging measures the natural gamma activity of the formation around the borehole. Gamma emissions generally can be correlated with rock type or fracture infilling. Deviations in the gamma log indicate changes in lithology or the presence of altered zones or mineralized fractures. Because the gamma log does not have a unique lithologic response, interpretation must be guided by the information provided by other logs.

Changes in the total dissolved solids (TDS) content of the water in the well are measured by use of specific-conductance logging. As the tool is lowered down the well, fluid resistivity is measured directly and its inverse, specific conductance, is calculated. To ensure that an undisturbed distribution of TDS was measured, fluid-resistivity logging was done first at each well. Because fluid resistivity generally is low in water containing salts, calibration of the fluid-resistivity tool can be difficult. Consequently, changes in specific conductance with depth are expected to provide more information than absolute values of specific conductance.

Temperature logging is used to sense locations where water enters or exits a well. In the absence of fluid flow in the borehole, water temperature generally decreases about 1 °F per 100 ft of depth (Keys, 1990). Deviations in this expected gradient indicate transmissive zones in the borehole.

Optical-televiewer (OTV) logging records a high-resolution, continuous, magnetically oriented, digitized 360-degree (°) color image of the borehole wall. The image permits direct examination of the borehole for locations and details of fractures, changes in lithology, water level, bottom of casing, and borehole enlargements. Because the images are magnetically oriented, they can be used to infer strike and dip of the rocks. Because the OTV is an optical system, however, it makes low-contrast features, such as small fractures in dark rocks, difficult to discern. Constituents in the water, such as sediment or oxidation products, can obscure the image.

Acoustic-televiewer (ATV) logging records a high-resolution, continuous, magnetically oriented image that is used to map the location and orientation of fractures that intersect a borehole. An acoustic beam is transmitted by the logging tool and the reflected wave is recorded. Fractures and other features in a borehole tend to scatter the acoustic wave, thereby producing high contrast on the borehole image.

A heat-pulse flowmeter (HPFM), sensitive to flows as low as 0.01 ± 0.005 gal/min, was used to measure the direction and rate of vertical flow in the boreholes under ambient and stressed conditions. Under ambient conditions, differences in hydraulic head between two transmissive fractures produce vertical flow in the borehole. If the hydraulic heads are nearly the same, however, no flow may be detected. Therefore, the HPFM also is used under pumping conditions to identify transmissive zones with similar ambient heads that would not be identified without stressing the aquifer. The HPFM uses a heat-pulse tracer that moves upward or downward with the flow in the borehole. After the tools described above were deployed to identify the depths of the fractured zones, the HPFM was used above and below the fractured zones to determine the transmissivity of the fractures and the direction of flow.

Most geophysical logging was done at rates of 11 to 15 feet per minute (ft/min). Because of the large amounts of data that the televiewers collected, their rates were about 8 ft/min. The HPFM was stationary when it was used to collect flow-rate information at specified depths.

Quality Assurance

The USGS maintained high-quality data by use of calibration checks before, during, and after deployment. All water-quality instruments (specific conductance, temperature, and water-quality probes) were shown to be functioning within manufacturers' specifications. Performance of water-level transducers was checked during site visits by use of standard tape-down techniques. During the data-collection period, little drift (no more than 0.04 ft) was found in any transducer. Water-level, specific conductance, and temperature data were subject to standard record review at the end of each water year; erroneous data (usually associated with normal maintenance at the wells) were removed from the published record.

Water Levels and Specific Conductance in Wells

Evidence for the role of fractures on the presence and movement of road-salt constituents, particularly chloride, in bedrock aquifers is presented in the following sections. Discussion of findings on a site-by-site basis also is included. Final daily-value data collected at the four study wells during water years 2008–9 are presented through the USGS Web portal (U.S. Geological Survey, 2009a) and in appendix 2.

Relation Between Specific Conductance and Chloride Concentrations

To reduce analytical costs and facilitate real-time reporting of data, USGS used specific conductance as a surrogate for chloride concentration on the basis of samples collected and analyzed for both constituents. Although this relation was not analyzed in a rigorous manner (as, for example, in Granato and Smith, 1999), the data showed a linear correlation over a range of about 3,000 μS/cm in six samples from three of the wells (fig. 10).

For the remainder of this report, changes in concentrations of specific conductance are considered indicative of changes in concentration of chloride. The following relation was established, with $R^2 = 0.9965$:

$$\text{Chloride concentration (in mg/L)} = \text{Specific conductance (in μS/cm)} \times 0.324 - 66, \quad (1)$$

which does not differ greatly from that used by Harte and Trowbridge (2010) for a small watershed in southern New Hampshire.

Site-by-Site Characteristics

Features of the geophysical, water-level, and specific-conductance data collected at each of the four wells are presented in this section. In some cases, the data are interpreted in the context of the landscape and (or) geologic environment at the site. Typical 10-day unit-value hydrographs of water levels in the four wells are presented in figure 11.

Gray

Active fractures, which are those that were actively contributing flow to the well under both static and pumped conditions, were found at depths of 84, 114, 186, and 294 ft below LSD (appendix 1). Under ambient conditions, water entered the borehole at the 294 ft depth and exited the borehole at the 114- and 186-ft depths. Vertical profiling (fig. 12) indicated slight water-chemistry changes near the bottom of casing (75 ft), probably in relation to the fracture at 84 ft, and larger changes below the 294 ft deep fracture. The specific conductance and temperature probe was set below the 84-ft fracture.

Water levels in the Gray well, collected between December 2007 and May 2009, exhibited consistent semidiurnal Earth-tide effects, fluctuating 0.1 ft to 0.2 ft daily (fig. 11). Daily water levels ranged from 10.26 to 13.96 ft below LSD during the period of record (fig. 2–1). Highest and lowest water levels were observed in April 2009 and July 2008, respectively. Water levels increased during spring snowmelt (from February through early April, both years) and declined through the summer months.

The minimum specific conductance, about 450 μS/cm, corresponds to a chloride concentration of about 80 mg/L, which is about four times higher than the background

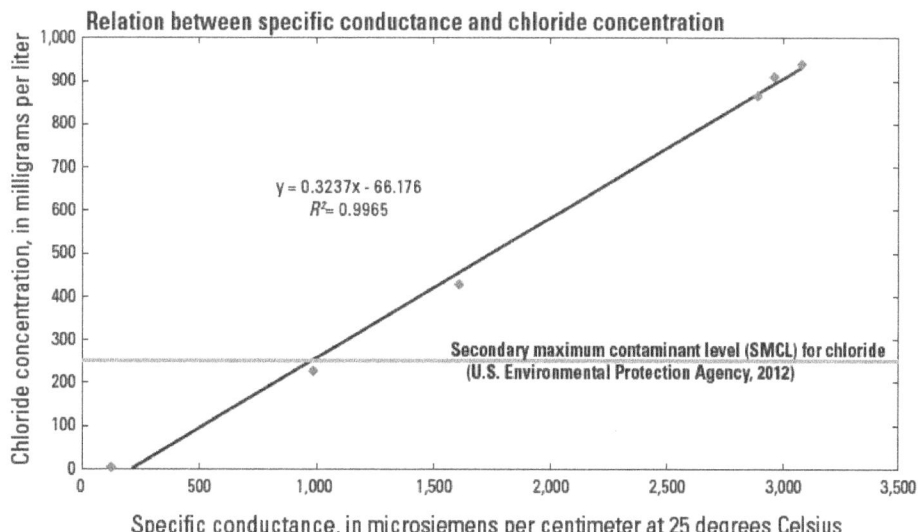

Figure 10. Relation between concentrations of chloride and specific conductance in three road-salt study wells, Maine.

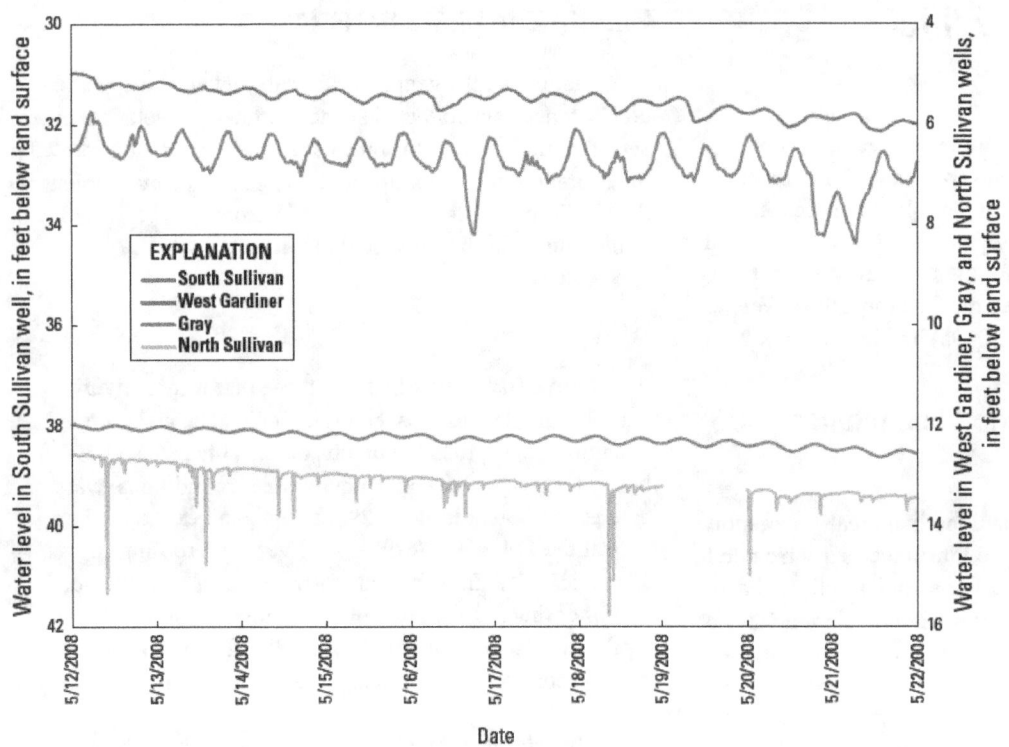

Figure 11.   10-day unit-value water levels for May 12–22, 2008, in four road-salt study wells, Maine.

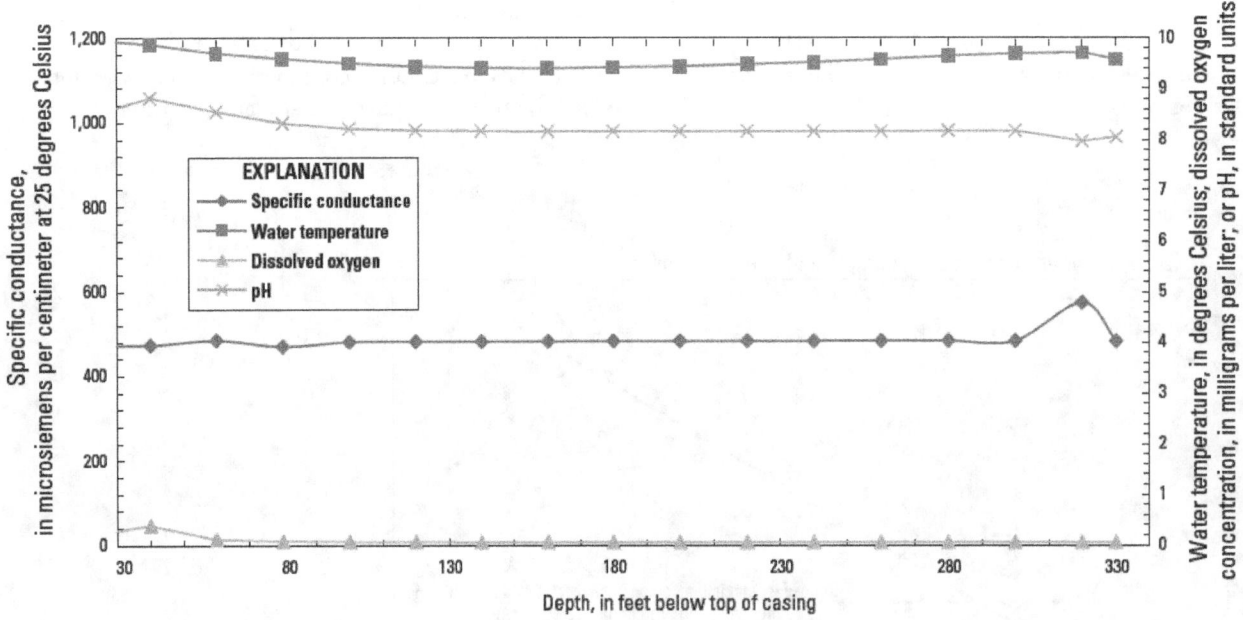

Figure 12.   Vertical water-quality profile in U.S. Geological Survey well 435103070191701 in Gray, Maine, October 22, 2008.

concentration reported by Olson and others (1985). Changes in specific conductance were neither large nor abrupt. Throughout the monitoring period, the range of recorded specific conductance was 450 to 485 µS/cm. Specific conductance increased from about 450 to 481 µS/cm from May 30 to June 28, 2008; the timing of this increase seems to indicate that the arrival of chloride to the well lags about 3 months behind spring snowmelt, which could be indicative of a long flow path for chloride. The peak conductance value may have been affected somewhat by the removal of the instruments from the well June 23–24 for geophysical logging.

From July 10 to October 19, 2008, daily water levels and specific conductance showed strong negative correlation (R^2 = 0.675) (fig. 13); that is, as water level rose during three consecutive events, specific conductance increased. This relation indicates that during storm-related recharge events, a net transport of salts occurs into the part of the aquifer penetrated by the Gray well. Salt is apparently stored along the flow path and is subsequently mobilized during recharge events. In general, peaks in specific conductance lagged about 2 days after peaks in water levels during these summer and autumn months.

Two mechanisms may have been involved in the long- and short-term responses of water levels and specific conductance in the Gray well. Long-term recharge, driven primarily by spring thaw and other wet periods over relatively long flow paths from land surface to the deep fractures in the well, provide a delayed response of specific conductance to changes in water level, whereas short-term changes in water level, driven primarily by summer or autumn rains when water levels are low, provide smaller pulses of high conductance water to the well. This could indicate that chloride is "pushed" into the well from surrounding fractures during rainfall events.

West Gardiner

Active fractures were found at depths of 36, 92, 120, 136, 170, 248, and 340 ft below LSD. Under ambient conditions, groundwater enters the well primarily at the 248 ft depth and exits at the 92 ft depth.

In general, water levels in the West Gardiner well were relatively higher than those in the other wells (fig. 2–2). Effects of pumping in the well were evident during the first few days of monitoring (after which MaineDOT moved the source of domestic supply to another well) and also during the middle of November, 2007, when geophysical logging was done at the well. The normal range of water levels was from 4 to 10.5 ft below LSD. Water levels increased during spring snowmelt (from February through early April, both years) and declined through the summer and autumn months. Water levels rose after about June 15, 2009, in response to the wetness of the summer. Sharp declines in water levels were probably in response to pumping in nearby wells. Water levels responded quickly to recharge events (fig. 2–2) and showed strong semi-diurnal Earth-tide effects (fig. 11).

Two extreme water levels were affected by actions being performed on a well 30 ft away (15-minute data not shown).

The lowest water level recorded, 19.93 ft below LSD on September 2, 2007, was in response to the nearby well being pumped; and the highest water level recorded, 3.90 ft below LSD on December 4, 2007, was in response to the same well being filled with bentonite as it was abandoned. The rapid and abrupt response of water levels to displacement of groundwater in the nearby well is indicative of the nature of groundwater flow in fractured-rock environments. It is likely that the same transmissive fracture intersected both the West Gardiner well and the nearby abandoned well.

Specific conductance generally was higher in the West Gardiner well than in other study wells, ranging from 681 to 3,890 µS/cm (corresponding to chloride concentrations of about 150 to 1,200 mg/L). The high chloride concentrations may have been associated historically with the presence of salt piles (which were covered in 1988) at the MaineDOT and MTA maintenance lots, as well as with ongoing winter maintenance practices at the MTA tollbooth area east of the well. The high chloride concentrations necessitated the abandonment of several wells nearby (Joshua Katz, Maine Department of Transportation, oral commun., 2007). Overall, specific conductance in the well shows a 2-year increase punctuated by (possibly) event-based peaks (fig. 2–2).

Figure 2–2 indicates that, after about May 1, 2008, specific conductance and water level shared an inverse relation; as water levels declined, specific conductance increased. Two explanations are possible. The first possible explanation for this inverse relation is that near-well recharge diluted the high concentration of chloride in the well. The probable ambient groundwater flow gradient in this area is from southeast to northwest, which tends to transport road salts applied to State Routes 9/126 away from the West Gardiner well rather than toward it. Thus, recharging groundwater probably is not transporting new chloride near the well to the subsurface. However, if dilution were occurring, specific conductance in the well should be decreasing consistently, but it is not (fig. 2–2).

The second possible explanation for the observed behavior of water levels and specific conductance is that a 3- to 6-month lag exists between precipitation events and the arrival of chloride to the well from a different source—the Maine Turnpike (Interstate 95), which lies upgradient toward the southeast. Some ancillary evidence for this hypothesis is on file with MaineDOT (Joshua Katz, Maine Department of Transportation, oral commun., 2008).

Vertical profiles from January 16, 2008, in the West Gardiner well (fig. 14) present evidence of the roles of fractures on all water-quality constituents. Dissolved oxygen concentration, pH, and specific conductance were approximately constant throughout the profile to a depth of 220 ft; at 220 ft, specific conductance increased to more than 5,000 µS/cm, pH decreased, and dissolved oxygen increased. At a depth of about 320 ft, specific conductance increased to 15,000 µS/cm (corresponding to a chloride concentration of 4,800 mg/L, about one-fourth that of seawater), pH decreased again, and dissolved oxygen content increased to concentrations observed far more regularly in surface water than in groundwater (Hem,

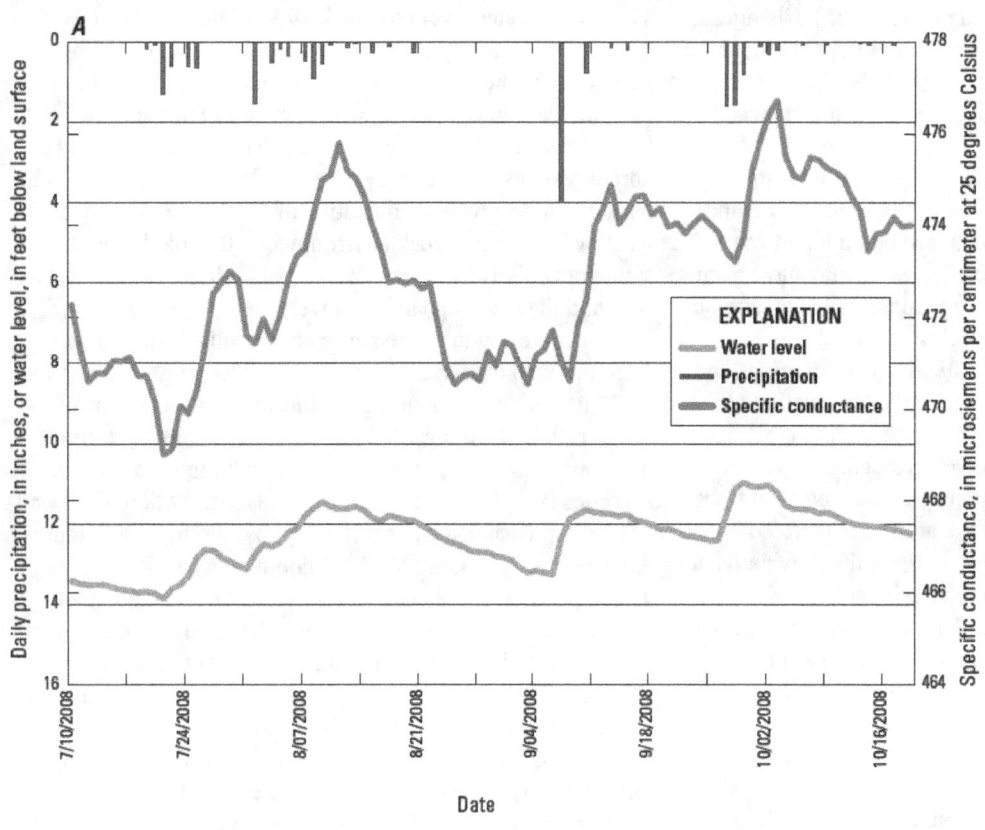

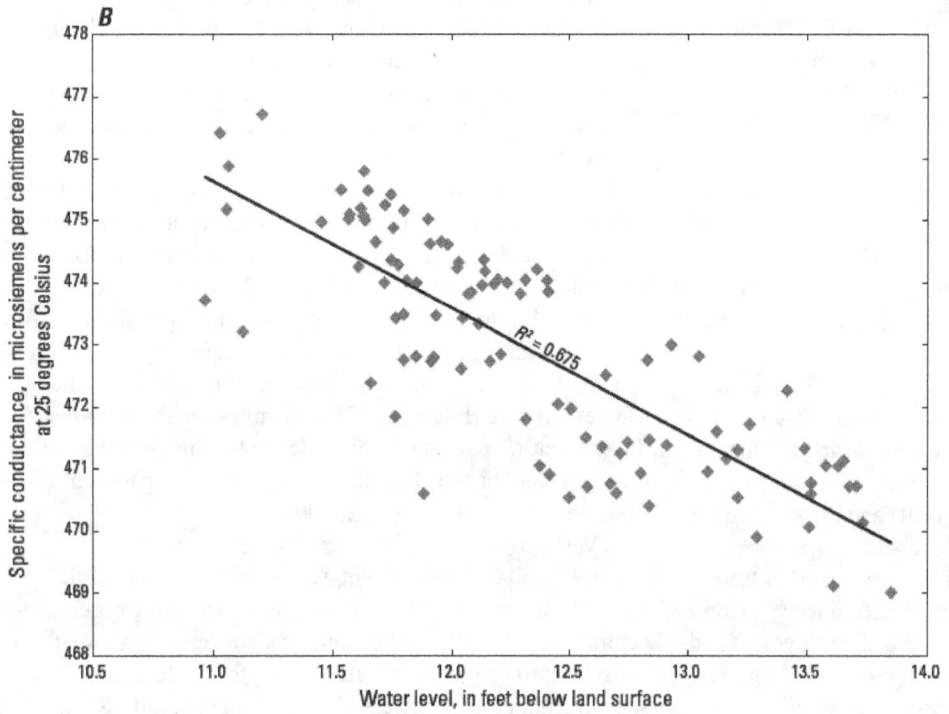

Figure 13. *A*, A hydrograph and *B*, correlation of daily water levels and specific conductance in U.S. Geological Survey well 435103070191701 in Gray, Maine, July 10–October 19, 2008.

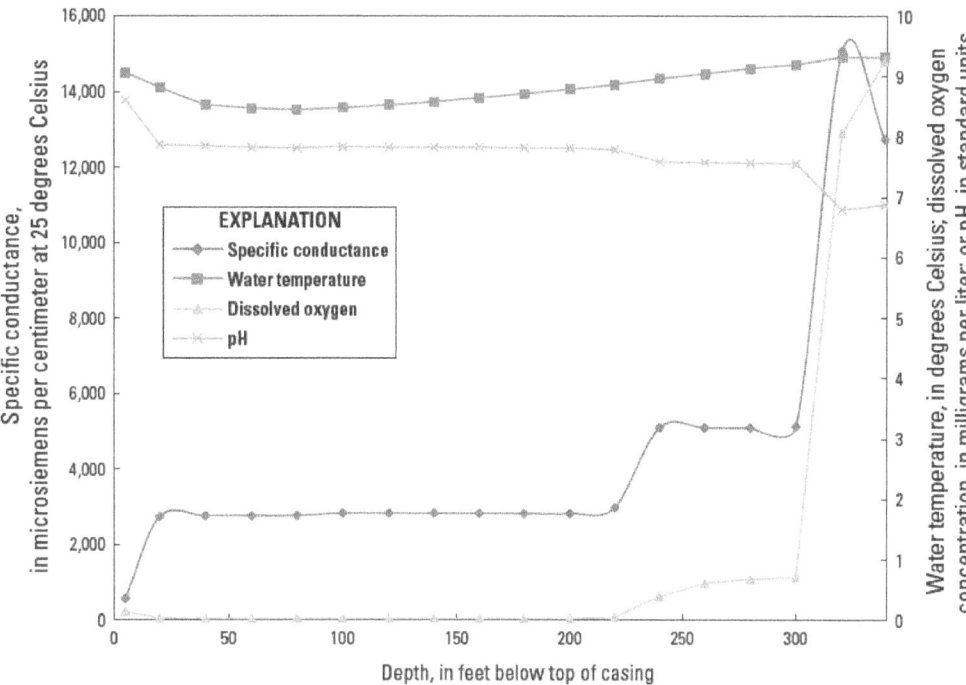

Figure 14. Vertical water-quality profile in U.S. Geological Survey well 441226069502201 in West Gardiner, Maine, January 16, 2008.

1985). These trends with depth also were observed in two subsequent profiles on different dates. One possible explanation is the existence of a near-vertical fracture that quickly transports oxygen-rich recharge water to the bottom of the well; if this were the case, however, it is likely that dilution effects on specific conductance would be more apparent.

North Sullivan

Geophysical data indicate that active fractures exist 15 ft below LSD at the base of the casing (table 1; appendix 1), and at depths of 25, 72, and 85 ft. Under ambient conditions, water entered the borehole at the 25-ft depth and exited the borehole at the 85-ft depth. Under pumping conditions, water entered the borehole at the 72-ft depth also.

Water levels in the North Sullivan well were affected by pumping from the well throughout the duration of the project. No Earth-tide effects were observed during any time period (fig. 11), even when the well was unused for a month. This well was relatively shallow (97 ft) and probably unconfined; that is, the water table is exposed to atmospheric pressure and is not subject to Earth tides.

The maximum and minimum water levels observed were 9.96 ft below LSD (April 7, 2009, during spring melt) and 17.89 ft below LSD (October 8, 2007, in response to pumping), respectively (15-minute values not shown). Water-level declines in response to pumping usually were in the range of 2 ft in 15-minute increments (fig. 11). Notwithstanding the effects of pumping, water levels generally increased during spring snowmelt (from February through early April) and autumn rains and declined through the summer and (or) winter months (fig. 2–3).

Specific conductance data used from the North Sullivan well in this report are on the basis of observations at the 80-ft depth. Although a second specific conductance probe was deployed at the 40-ft depth in June 2008, the two sets of specific-conductance data did not vary greatly ($R^2 = 0.95$) after about July 1, 2008.

Initially, in autumn 2007 and winter 2007–8, specific conductance decreased in response to increases in water level from recharge (fig. 15), probably because the first recharge water entered the well through large fractures that did not contain appreciable concentrations of salt; chloride in the well was diluted initially from about 1,000 to 900 µS/cm (about 250–225 mg/L chloride). Thereafter, however, recharge water brought additional chloride with it, probably through small fractures or through the bulk matrix, and specific conductance increased with water level. Several examples of the initial decrease, then rapid increase in specific conductance with recharge are shown in figure 15 for December 24–25, 2007, and February 13 and April 30, 2008.

The degree to which specific conductance responded to recharge varied on the basis of antecedent conditions. About 26 in. of snow fell between December 24, 2007, and January 5, 2008, resulting in a snowpack depth of about 2 ft deep; temperatures were between 38 and -10 °F (National Oceanic and Atmospheric Administration, 2008). On January 9, 2008, water levels began to rise at noon in response to a major thaw January 6–13 (maximum high temperature of 52 °F on January 8 and minimum low temperatures above freezing January 7–9), during which no precipitation fell; temperature and specific conductance responded about 16 hours later. In total, the January 6–13 thaw caused the snowpack to diminish to 8 in., water levels to rise about 1.5

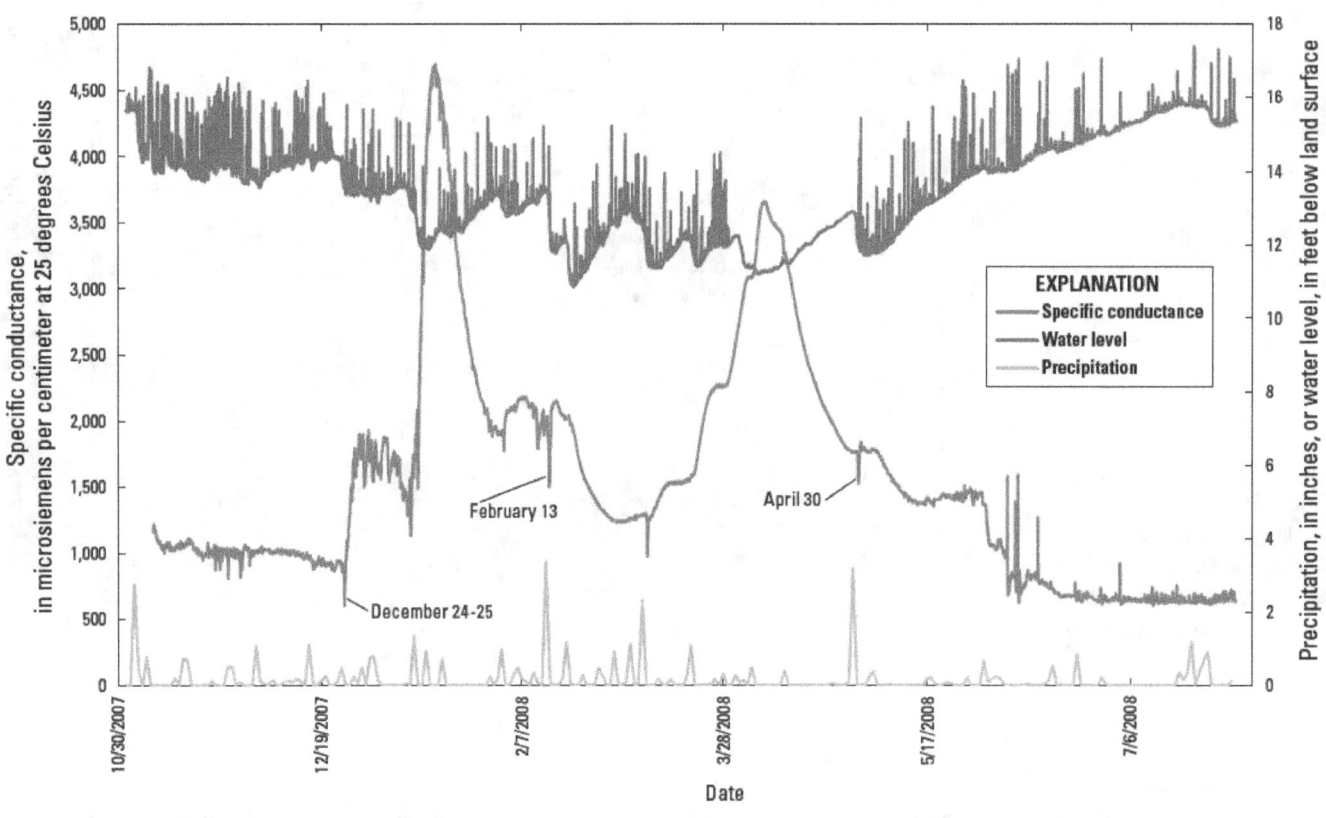

Figure 15. Water levels and specific conductance in U.S. Geological Survey well 443145068131801 in North Sullivan, Maine, November 2007–July 2008.

ft, water temperature to rise about 2 °F, and specific conductance to rise about 3,000 µS/cm. Increases in water levels and specific conductance also occurred around January 30 in response to warm temperatures January 29–30 (maximum high temperature 45 °F on January 30 and minimum low temperatures slightly below freezing; decrease in snowpack); and on February 14 in response to warm temperatures February 13–15 (maximum high temperature 43 °F on February 13 and minimum low temperatures 15–18 °F). The February 13–15 thaw was accompanied by more than 5 in. of new snow but produced an overall decrease in the snowpack. A warm period February 18–19 (maximum temperature 48 °F on February 18, and minimum low temperatures around freezing) produced a decrease in snowpack, an increase in water levels, and little change in temperature or specific conductance. Between February 15 and 18, less than 0.5 in. of snow had fallen, so additional road salt probably had not been applied. In all likelihood, the January 9–10 thaw allowed an accumulation of road salts that had been applied during the antecedent snowy weather to enter the well with melted snow, as evidenced by the large spike in specific conductance that lasted about 3 weeks; the later events produced smaller spikes in specific conductance because antecedent weather had required the use of less road salt.

The effects of the annual spring snowmelt were observed in March and April 2008. Snowpack depth decreased or held steady throughout March and disappeared around April 7. The highest water level since the thaw of February 18 occurred on April 5. Specific conductance rose steadily from March 21 through April 7, 2008, probably in response to the annual spring snowmelt; at the same time, temperature decreased 1 °F because the snowmelt water was colder than the ambient water.

From June 6 through 9, 2008, when water levels spiked downward (3-ft declines) due to pumping, specific conductance spiked upward as much as 700 µS/cm. Later in summer 2008, on many occasions when water levels spiked downward due to pumping, specific conductance did not respond. These data, in concert with those described above, indicate that hydrologic events, and in particular their timing in relation to the presence of road salt, played a larger role in the transport of chlorides than did pumping alone.

South Sullivan

Geophysical data indicate that active fractures exist at depths of about 40, 53, 186, 198, and 208 ft below LSD.

Under ambient conditions, water enters the borehole at the 186- and 198-ft depths and exits the borehole at the 40- and 53-ft depths. The 208-ft fracture became active under pumping conditions.

Water levels in the South Sullivan well exhibited strong semi-diurnal fluctuations of 1 foot or more (fig. 11), possibly in response to both Earth tides and ocean tides. Occasional, periodic declines of water levels as much as 2 ft (fig. 11) may have been in response to pumping from a nearby well that was not identified. The minimum and maximum water levels recorded were 28.06 ft below LSD and 37.22 ft below LSD, respectively (15-minute data not shown). Water levels generally increased during spring snowmelt (from February through early April) and declined through the summer months (fig. 2–4).

At the 220 ft depth, specific conductance remained above about 750 μS/cm (corresponding to 180 mg/L chloride) throughout the year (fig. 2–4). Water-level fluctuations had no major effect on specific conductance at 220 ft, which is below the active fractures. Small changes in specific conductance at 42 ft, however, lagged about 5 days behind changes in water level. The South Sullivan well was slightly uphill from U.S. Route 1; runoff and infiltration probably were directed away from the well head, and it is likely that the topography of the area masked any possible cause-and-effect relation that might otherwise have been observed.

Effects of Stresses on Chloride Concentrations

The data presented in the "Site-by-Site Characteristics" section indicate that natural and anthropogenic hydrologic stresses play meaningful roles in the presence of chloride in groundwater. Natural stresses, including recharge by precipitation and thawing, generally affect chloride concentrations by transporting chloride to the well from overlying soil and rock. Anthropogenic stress (pumping) affects chloride concentrations by creating artificially steep, temporary gradients between the water table and the water level in the well. Effects of stresses are seen throughout the year.

Precipitation

Specific conductance in the Gray and North Sullivan wells responded positively to recharge events throughout the year. In general, when water levels rose in these wells, specific conductance increased, although to varying degrees. Daily-value water levels and specific conductance from the Gray well were correlated (fig. 13B), and the data indicated that changes in specific conductance lagged about 3 days behind changes in water levels. This lag may have been because the initial water-level changes were because of increased pressure from the water column, and not from an actual influx of recharging water; or it may have been because the specific conductance probe was set at 190 ft and the chloride entering the well with recharge water required 3 days to diffuse to

that depth. Recharge-related effects were most marked in the North Sullivan well, probably because of its shallowness, the shortness of the well casing (15 ft below LSD), and pumping in the well.

The positive response of specific conductance to changes in water level indicates that the recharge water typically was higher in chlorides than the water in the well—not just during winter thaws or spring recharge, but throughout the year. In general, after early recharge of relatively clean water, specific conductance began to rise, indicating that the recharge water percolating through smaller fractures or the bulk matrix contained higher concentrations of chlorides than did the early recharge water. These hydrologic responses imply that chlorides are retained in the unsaturated zone, small fractures, and (or) the bulk matrix throughout the year, not just during the winter.

The responses to recharge observed in the Gray and North Sullivan wells were not observed in the West Gardiner well, where the conceptual flow model indicates that short-term recharge near the well tends to dilute chloride present in the well, whereas long-term recharge along flow paths from the southeast tend to concentrate chlorides, even when groundwater level declines.

Thawing

Specific conductance in the Gray and North Sullivan wells showed a notable response to snowmelt. Although recharge can occur throughout the winter (as evidenced by increases in water levels in all wells during winter 2007–8 (appendix 2), the largest values of specific conductance—or increases in concentration of chloride—were associated with thaws. Labadia and Buttle (1996) found that as much as 50 percent of the applied road salt remains in the snowbank until a thaw; if a connecting fracture exists from the surface to the subsurface, concentrations or road-salt constituents will spike during a thaw. Once again, the most remarkable example was the North Sullivan well.

Pumping

The combination of effects observed at the North Sullivan well during summer 2008 supports the hypothesis that chloride is sequestered in the matrix or disconnected fractures until mobilized by recharge. During the late spring, when ambient water levels and soil moisture were presumably high from the spring thaw, specific conductance increased in response to pumping; chloride-bearing groundwater was drawn into the well. During the late summer, when ambient water levels and soil moisture were presumably low from the effects of evapotranspiration, specific conductance did not respond to pumping; groundwater drawn to the well did not contain new concentrations of chloride.

These data support the idea that chloride is "suspended" in the bulk matrix or in small, partially disconnected fractures until activated by recharge. When hydrologic events were few and far between, specific conductance was unlikely to respond to pumping alone; but when precipitation was sufficient to produce recharge or when air temperature was sufficient to produce significant thawing, chloride was transported into the well, causing an increase in specific conductance.

Effects of Fractures on Water Quality

Water-quality profiles of the wells supported data acquired by geophysical methods and confirmed that the active fractures in each well played prominent roles in the quality of water in the well. Distinct changes in water chemistry were observed at depths correlating to the presence of active fractures, as, for example, in the West Gardiner well (fig. 14). Profiles in the South Sullivan well (not shown) were similar, though not as dramatic.

The behavior of dissolved oxygen concentrations with depth in the West Gardiner well (fig. 14) are probably the most noteworthy example of the role of fractures on water quality. From land surface to a depth of 220 ft, dissolved oxygen concentration decreased from about 0.1 to 0.03 mg/L, which is typical for groundwater. From 220 to 300 ft, dissolved oxygen concentration increased to about 0.7 mg/L, and below 300 ft, dissolved oxygen concentration increased to 10 mg/L. This behavior was repeated during three separate visits, and the meter was found to be within calibration specifications each visit after the profiling effort. One possible explanation is the existence of a near-vertical fracture that quickly transports oxygen-rich recharge water to the bottom of the well; if this is true, then the same fracture can readily be cited as playing a major role in the transport of chloride to the subsurface as well.

Summary and Conclusions

In 2007, in cooperation with Maine Department of Transportation (MaineDOT), the U.S. Geological Survey began an investigation of the processes that enable chlorides from road salt to be present in bedrock aquifers. Data collected by MaineDOT's Well Claims Unit had shown that chloride concentrations in wells affected by road salt increased in the autumn, after water levels declined in the summer, and before road salts were applied in the winter. This investigation sought to determine a relation between specific conductance and chloride concentration, and, on the basis of that relation, to explore the role of fractures on the quality of groundwater; a link between road salting and chloride contamination; and the hydrologic conditions and timing of events that affected chloride concentrations in groundwater.

Continuous data on water levels, specific conductance, and temperature were collected from 2007 through 2009 at four bedrock wells in Maine that were known to be affected by road salts. Geophysical data were collected to locate fractures

and determine whether the fractures were contributing groundwater to the well under static and pumped conditions. Vertical profiles of water-quality indicators (dissolved oxygen, specific conductance, pH, and temperature) were collected at least once at the three wells not being used for domestic supply. Water-quality samples were collected to establish a linear relation between specific conductance and chloride concentration.

Geophysical methods identified the active fractures in each well, and vertical profiling verified the contributions of these fractures to the distributions of water quality with depth in each well. All abrupt changes in water-quality constituents during profiling were at depths corresponding to those of the active fractures identified by geophysical methods.

Data collected during the project indicated that chloride concentration was relatively stable in the wells, except when concentrations increased during the main spring recharge or during major winter thaws. Minimum specific conductance, about 450 µS/cm in the Gray well (corresponding to about 80 mg/L chloride), indicated that chlorides at concentrations greater than background levels continued to be present after large recharge events such as snowmelt and rainfall, especially in the spring. In contrast, in highly permeable, unconsolidated environments, the literature documents a breakthrough-curve behavior for chloride.

Contamination of the wells was linked to road salting in at least one well on the basis of large increases in specific conductance associated with winter thaws and spring recharge, when road salts are most prevalent in the environment. A winter thaw in January 2008 produced the largest spike in specific conductance observed at the North Sullivan well, and specific conductance did not recover to the pre-thaw concentration for 3 weeks. Annual high specific conductances were observed in at least two wells during the spring snowmelt.

Data showed that chloride can be transported to the wells throughout the year, not only during the winter. Although the largest changes in specific conductance were associated with thaws, other hydrologic events also triggered changes in specific conductance. Recharge events throughout the summer, for example, produced increases in specific conductance. These data indicate that road salts are sequestered in the bulk matrix of the unsaturated zone or in isolated fractures, and that, when recharge events occur, they can be flushed into the system.

Hydrologic events were the driving factor behind changes in specific conductance. Winter and spring thaws and rainfall transported chloride to the groundwater during recharge events. Some effects of pumping on chloride concentrations also were observed, but substantially greater changes in chloride concentration were observed when recharge was a factor than when pumping was the only factor.

The magnitude of chloride concentrations with depth in the wells also was related to the presence of fractures. Vertical profiles consistently showed relatively abrupt changes in water-quality constituents at depths of active fractures.

References Cited

Aichele, S., 2004, Arsenic, nitrate, and chloride in ground-water, Oakland County, Michigan: U.S. Geological Survey Fact Sheet 2004–3120, 6 p.

Andrews, W.J., 1996, Water-quality assessment of part of the Upper Mississippi River Basin, Minnesota and Wisconsin—Ground-water quality in an urban part of the Twin Cities metropolitan area, Minnesota: U.S. Geological Survey Water-Resources Investigations Report 97–4248, 54 p.

Andrews, W.J., Stark, J.R., Fong, A.L., and Fallon, J.D., 2005, Water-quality assessment of part of the upper Mississippi River basin, Minnesota and Wisconsin—Ground-water quality along a flow system in the Twin Cities metropolitan area, Minnesota, 1997–98: U.S. Geological Survey Scientific Investigations Report 2005–5120, 44 p.

Bester, M.L., Frind, E.O., Molson, J.W., and Rudolph, D.L., 2006, Numerical investigation of road salt impact on an urban watershed: Ground Water, v. 44, no. 2, p. 165–175.

Blasius, B.J., and Merritt, R.W., 2002, Field and laboratory investigations on the effects of road salt (NaCl) on stream macroinvertebrate communities: Environmental Pollution, v. 120, no. 2, p. 219–231.

Blomkvist, G., and Johanssen, E.-L., 1999, Airborne spreading and deposition of de-icing salt—A case study: The Science of the Total Environment, v. 235, p. 161–168.

Caswell, W.B., 1979, Maine's ground-water situation: Ground Water, v. 17, no. 3, p. 235–243.

Corsi, S.R., Graczyk, D.J., Geis, S.W., Booth, N.L., and Richards, K.D., 2010, A fresh look at road salt—Aquatic toxicity and water-quality impacts on local, regional, and national scales: Environmental Science and Technology, v. 44, no. 19, p. 7376–7382.

Cunningham, W.L., and Schalk, C.W., comps., 2011, Ground-water technical procedures of the U.S. Geological Survey: U.S. Geological Survey Techniques and Methods, book 1, chap. A1, 151 p., at http://pubs.usgs.gov/tm/1a1/.

Demers, C.L., and Sage, R.W., 1990, Effects of road deicing salt on chloride levels in four Adirondack streams: Water, Air, and Soil Pollution, v. 49, no. 3–4, p. 369–373.

Fobes, C.B., 1942, Snowfall in Maine: Geographical Review, v. 32, no. 2, p. 245–251.

Foos, A., 2003, Spatial distribution of road salt contamination of natural springs and seeps, Cuyahoga Falls, Ohio, USA: Environmental Geology, v. 44, p. 14–19.

Forman, R.T., and Deblinger, R.D., 2000, The ecological road-effect zone of a Massachusetts (U.S.A.) suburban highway: Conservation Biology, v. 14, no. 1, p. 36–46.

Godwin, K.S., Hafner, S.D., and Buff, M.F., 2003, Long-term trends in sodium and chloride in the Mohawk River, New York—The effect of fifty years of road-salt application: Environmental Pollution, v. 124, p. 273–281.

Granato, G.E., Church, P.E., and Stone, V.J., 1995, Mobilization of major and trace constituents of highway runoff in groundwater potentially caused by deicing chemical migration: Washington, D.C., Transportation Research Record 1483, p. 92–104.

Granato, G.E., and Smith, K.P., 1999, Estimating concentrations of road-salt constituents in highway-runoff from measurements of specific conductance: U.S. Geological Survey Water-Resources Investigations Report 99–4077, 22 p.

Harte, P.T., and Trowbridge, P.R., 2010, Mapping of road-salt-contaminated groundwater discharge and estimation of chloride load to a small stream in southern New Hampshire, USA: Journal of Hydrological Processes, v. 24, no. 17, p. 2349–2368, accessed April 16, 2012, at http://dx.doi.org/10.1002/hyp.7645.

Hem, J.D., 1985, Study and interpretation of the chemical characteristics of natural water: U.S. Geological Survey Water-Supply Paper 2254, 264 p.

Jackson, R.B., and Jobbagy, E.B., 2005, From icy roads to salty streams: Proceedings of the National Academy of Sciences, v. 102, no. 41, p. 14487–14488.

Johnson, C.D., Haeni, F.P., Lane, J.W., Jr., and White, E.A., 2002, Borehole-geophysical investigation of the University of Connecticut landfill, Storrs, Connecticut: U.S. Geological Survey Water-Resources Investigations Report 01–4033, 187 p.

Kaushal, S.S., Groffman, P.M., Likens, G.E., Belt, K.T., Stack, W.P., Kelly, V.R., Band, L.E., and Fisher, G.T., 2005, Increased salinization of fresh water in the northeastern United States: Proceedings of the National Academy of Sciences, v. 102, no. 38, p. 13517–13520.

Kelly, W.R., 2008, Long-term trends in chloride concentrations in shallow aquifers near Chicago: Ground Water, v. 46, no. 5, p. 772–781.

Keys, W.S., 1990, Borehole geophysics applied to ground-water investigations: U.S. Geological Survey Techniques of Water-Resources Investigations, book 2, chap. E–2, 149 p.

Kostick, D.S., 1994, Salt, *in* Metals and minerals: U.S. Bureau of Mines Minerals Yearbook 1992, v. 1, p. 1103–1132, accessed April 11, 2012, at http://digicoll.library.wisc.edu/cgi-bin/EcoNatRes/EcoNatRes-idx?type=article&did=EcoNatRes.MinYB1992v1.WBolen&id=EcoNatRes.MinYB1992v1&isize=M

Kostick, D.S., 1996–2012, Salt, *in* Metals and minerals: U.S. Geological Survey Minerals Yearbook 1994–2010, accessed April 11, 2012, at http://minerals.usgs.gov/minerals/pubs/commodity/salt/index.html.

Kunze, A.E., and Sroka, B.N., 2004, Effects of highway deicing chemicals on shallow unconsolidated aquifers in Ohio—Final report: U.S. Geological Survey Scientific Investigations Report 2004–5150, 187 p.

Labadia, C.F., and Buttle, J.M., 1996, Road salt accumulation in highway snow banks and transport through the unsaturated zone of the Oak Ridges moraine, southern Ontario: Hydrological Processes, v. 10, p. 1575–1589.

Maine Department of Transportation, 2004, Road salt contamination claims—Technical and legal issues: Maine Department of Transportation, accessed February 4, 2009, at http://www.maine.gov/mdot/community-programs/road-saltclaims_000.php.

Maine Interagency Report, 2001, Sand and salt storage in Maine—Report to the 120th Maine Legislature: Augusta, Maine, Maine Department of Environmental Protection and Maine Department of Transportation, January 6, 47 p.

Maine Legislature, 2012, Maine revised statute title 23—Highways: Maine Legislature, 331 p., available at http://www.mainelegislature.org/legis/statutes/23/title23.pdf.

Methuel, R.W., 2007, Effects of deicing salts on the chloride levels in water and soil adjacent to roadways: Michigan Department of Transportation Construction and Technology Division, Research Report R–1495, 16 p.

National Oceanic and Atmospheric Administration, 2002, Snowfall—Average total in inches: National Oceanic and Atmospheric Administration database, accessed September 29, 2010, at http://lwf ncdc.noaa.gov/oa/climate/online/ccd/snowfall.html.

National Oceanic and Atmospheric Administration, 2008, Daily climatological data for Ellsworth, Maine, January 2008: National Climatic Data Center data on file with U.S. Geological Survey, Augusta, Maine.

National Weather Service, 2011, Wet year in Portland, Maine: CNYCentral.com, accessed August 31, 2011, at http://www.cnycentral.com/weather/blog_post.aspx?id=398150#.Tl5Ep2HMo8k.

New Hampshire Department of Environmental Services, 2010, Sodium and chloride in drinking water: New Hampshire Department of Environmental Services Factsheet WD–DWGB–3–17, 3 p., accessed September 10, 2012, at http://des.nh.gov/organization/commissioner/pip/factsheets/dwgb/documents/dwgb-3-317.pdf.

Nimiroski, M.T., and Waldron, M.C., 2002, Sources of sodium and chloride in the Scituate Reservoir drainage basin, Rhode Island: U.S. Geological Survey Water-Resources Investigations Report 02–4149, 16 p.

Olson, A.C., Allenwood, M., and Williams, J.S., 1985, Replacement of salt contaminated water supplies in bedrock aquifers in Maine: Proceedings of the Second Annual Eastern Regional Ground Water Conference, National Ground Water Association, Ground Water Technology Division, July 16–18, 1985, Portland, Maine, p. 201–212.

Osberg, P.H., 1968, Stratigraphy, structural geology, and metamorphism of the Waterville-Vassalboro area, Maine: Maine Geological Survey Bulletin, no. 20, 64 p.

Osberg, P.H., Hussey, A.M., II, and Boone, G.M., 1985, Bedrock geologic map of Maine: Maine Geological Survey, 1 sheet, scale 1:500,000.

Ostendorf, D.W., Hinlein, E.S., Rotaru, C., and DeGroot, D.J., 2006, Contamination of groundwater by outdoor highway deicing agent storage: Journal of Hydrology, v. 326, no. 1–4, p. 109–121.

Ostendorf, D.W., Palmer, R.N., and Hinlein, E.S., 2009, Seasonally varying highway deicing agent contamination in a groundwater plume from an infiltration basin: Hydrology Research, v. 40, no. 6, p. 520–532.

Paschka, M.G., Ghosh, R.S., and Dzombak, D.A., 1999, Potential water-quality effects from iron cyanide anticaking agents in road salt: Water Environment Research, v. 71, p. 1235–1239.

Ramakrishna, D.M., and Viraraghavan, T., 2005, Environmental impact of chemical deicers—A review: Water, Air, and Soil Pollution, v. 166, no. 1–4, p. 49–63.

Richburg, J.A., Patterson, W.A., and Lowenstein, F., 2001, Effects of road salt and *Phragmites australis* invasion on the vegetation of a western Massachusetts calcareous lake-basin fen: Wetlands, v. 21, no. 2, p. 247–255.

Rosenberry, D.O., Bukaveckas, P.A., Buso, D.C., Likens, G.E., Shapiro, A.M., and Winter, T.C., 1999, Movement of road salt to a small New Hampshire lake: Water, Air, and Soil Pollution, v. 109, p. 179–206.

Rubin, Jonathan, Gärder, P.E., Morris, C.E., Nichols, K.L., Peckenham, J.M., McKee, Peggy, Stern, Adam, and Johnson, T.O., 2010, Maine winter roads—Salt, safety, environment and cost: University of Maine, Margaret Chase Smith Policy Center, February, 142 p. (Also available at http://mcspolicycenter.umaine.edu/2010/02/19/winterroadmaint-final/.)

Snow, M.S., Kahl, J.S., Norton, S.A., and Olsen, Christine, 1990, Geochemical determination of salinity sources in ground water wells in Maine, *in* Focus Conference on Eastern Regional Ground Water, Springfield, Mass., Proceedings: Columbus, Ohio, National Ground Water Association, October 17–19, 1990, p. 313–327.

State of Maine Judicial Branch, 2008, Jeanannette Waning versus Department of Transportation, Maine Supreme Judicial Court docket Cum–07–651: State of Maine Judicial Branch, 8 p., accessed February 4, 2009, at http://www.courts.state.me.us/court_info/opinions/2008%20documents/08me95wa.pdf.

Stewart, D.B., 1998, Geology of northern Penobscot Bay, Maine, with contributions to geochronology by Robert D. Tucker: U.S. Geological Survey Miscellaneous Investigations Series Map I–2551, 2 sheets, map scale 1:62,500.

Thompson, W.B., and Borns, H.W., Jr., 1985, Surficial geologic map of Maine: Maine Geological Survey, scale 1:500,000.

Transportation Research Board, 1991, Highway deicing—Comparing salt and calcium magnesium acetate: Washington, D.C., National Research Council, Transportation Research Board Special Report 235, accessed December 17, 2008, at http://onlinepubs.trb.org/Onlinepubs/sr/sr235 html.

U.S. Environmental Protection Agency, 2012, Secondary drinking water regulations—guidance for nuisance chemicals: U.S. Environmental Protection Agency, accessed October 30, 2012, at http://water.epa.gov/drink/contaminants/secondarystandards.cfm.

U.S. Geological Survey, 2009a, National Water Information System (NWISWeb): U.S. Geological Survey database, accessed November 7, 2009, at http://waterdata.usgs.gov/me/nwis/.

U.S. Geological Survey, 2009b, Silurian Waterville Formation: U.S. Geological Survey database, accessed July 6, 2009, at http://tin.er.usgs.gov/geology/state/sgmc-unit.php?unit=MESw%3B0.

U.S. Geological Survey, 2010, Salt statistical compendium: U.S. Geological Survey, accessed April 11, 2012, at http://minerals.usgs.gov/minerals/pubs/commodity/salt/stat/index.html.

University of Maine, School of Law, 2003, Bouchard versus City of Lewiston civil action, Androscoggin Superior Court docket CV–03–124: University of Maine, School of Law, 9 p., accessed February 4, 2009, at http://mainelaw.maine.edu/library/SuperiorCourt/decisions/ANDcv-03-124.pdf.

Williams, D.D., Williams, N.E., and Cao, Y., 1999, Road salt contamination of groundwater in a major metropolitan area and development of a biological index to monitor its impact: Water Research, v. 34, no. 1, p. 127–138.

THIS PAGE INTENTIONALLY LEFT BLANK

Appendix 1.   Geophysical Data

The following abbreviations and acronyms are used in appendix 1:

ABI	acoustic televiewer
amb	ambient
Azi	azimuth
cond	conductance
cps	counts per second
deg	degrees
deg C	degrees Celsius
deg F	degrees Fahrenheit
DOT	Department of Transportation
EMI	electromagnetic induction
Ft, ft, or FT	feet
gal/min	gallons per minute
HPFM	heat-pulse flowmeter
ID	identifier
in or in.	inches
LMP	log-measuring point
ME	Maine
MP	measuring point
MSI	Mount Sopris Instruments
OBI	optical televiewer
OGW–BG	Office of Groundwater Borehole Geophysics Unit
SC	specific conductance
SN	serial number
Temp	temperature
uS/cm	microsiemens per centimeter
USGS	U.S. Geological Survey
WL	water level

Company USGS   **Site ID** 435103070191701   **Station name** CW-2028

Other ID Gray well   **Date of log** 06/23/2008

County/State Cumberland/Maine   **Office/logging unit** OGW-BG Storrs

Logging operator MJN, CWS   **Observer**   CDJ

Description of log-measuring point (LMP) Top of casing

Height of LMP above/below land-surface datum (in feet) 1.0

Altitude of LMP   342 feet   **Magnetic declination**   ~-14

Borehole depth/diameter/type   327 feet / 6 in. / open

Casing depth/diameter/type   75.25 feet / 6 in. / steel

Borehole fluid type   Water

Borehole fluid depth and date   14.10 feet at 10:05 on 6/23/08

Hydrologic conditions   Ambient- Pre-injection conditions

Remarks   Stickup=1 foot, Pumping at 0.25-.3 gal/min was conducted as part of stressed flowmeter tests on 6/24/08.
ME DOT reports elevation of MP at 327.55, USGS reports 342 feet. Well reports indicate 20 feet of casing, but there appears to be casing to 75.25 feet

Logs in composite, date and time, manufacturer, serial number (SN), condition

EMI and Gamma, 06/23/08, MSI-2PIA+2PGA, SN: 2307, gamma factory calibrated; count per second used
EMI tool - calibrated after log; data adjusted

Fluid and Caliper, 06/23/08, MSI-2PCA+2FSB, SN: 3009, calibration checked after log; data adjusted to calibration values

OBI-40-Mk4,   06/24/08, MSI-ALT, MSI SN: 073612, Alt SN:061101, tilt and direction check at time of log

ABI-40,   06/23/08, MSI-ALT, MSI SN: 3078, Alt SN: 020906, tilt and direction check at time of log

HPFM,   06/24/08, MSI 2293, SN: 2060, calibration values updated in MsHeat for tool 2060
ambient and pumping logs collected

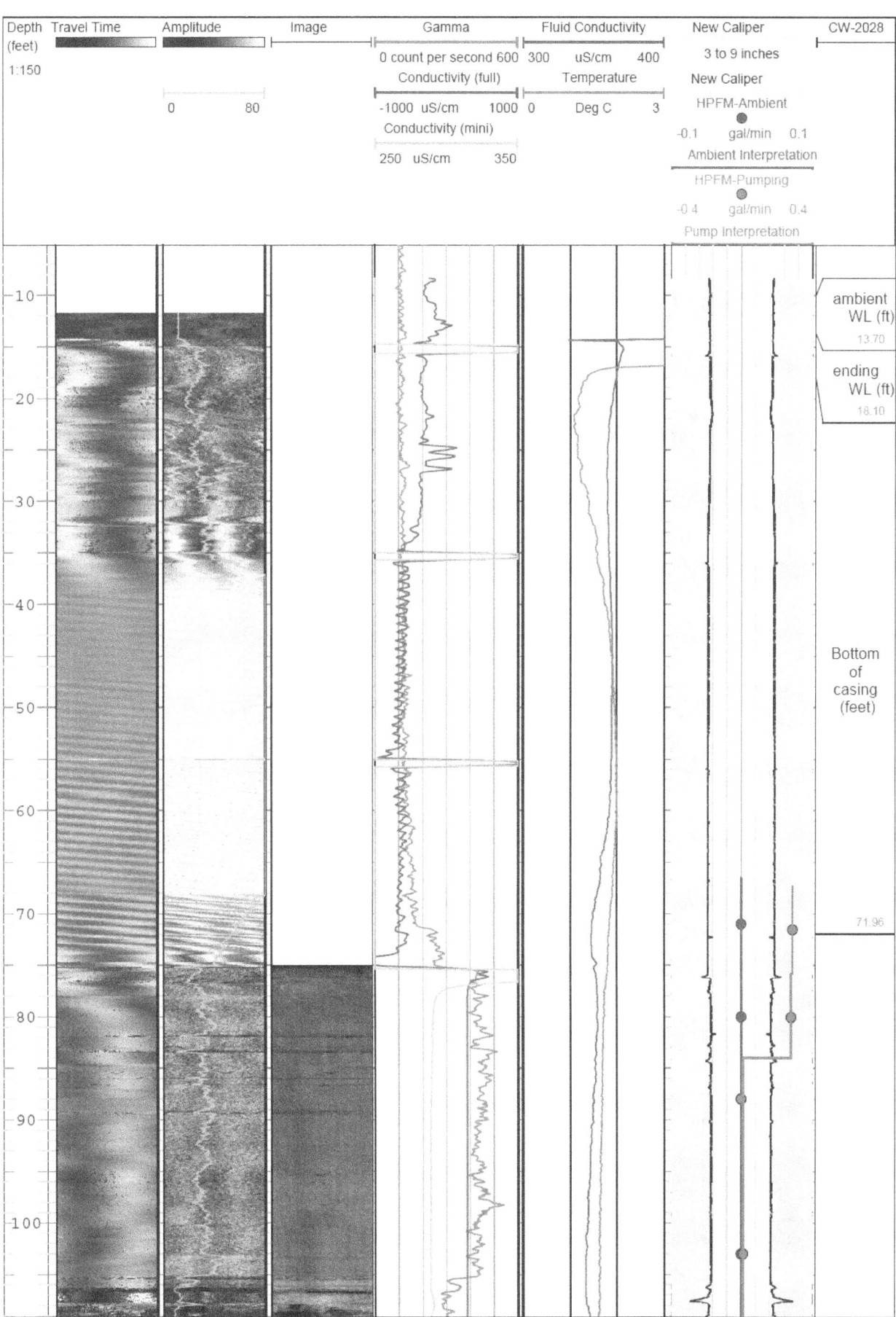

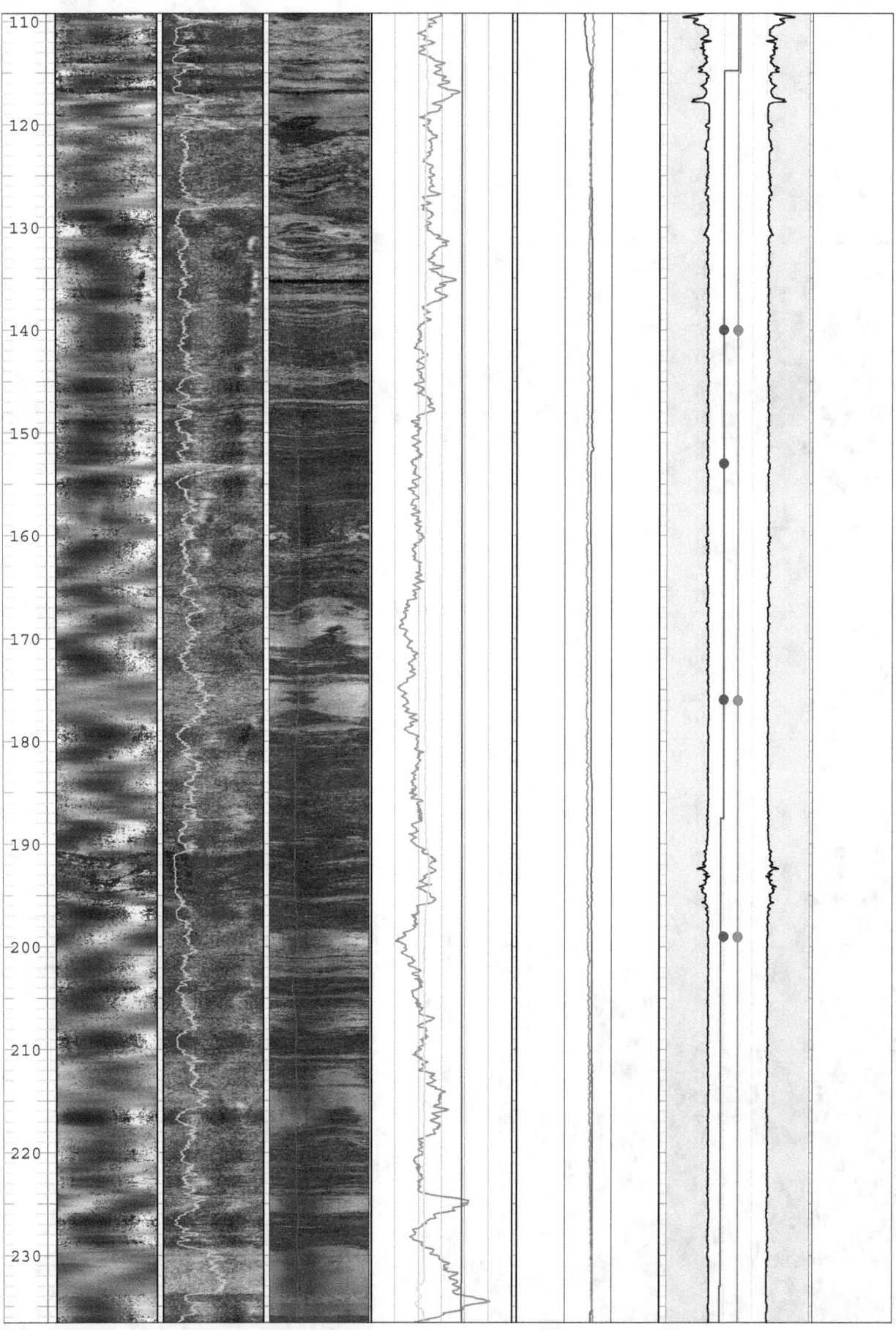

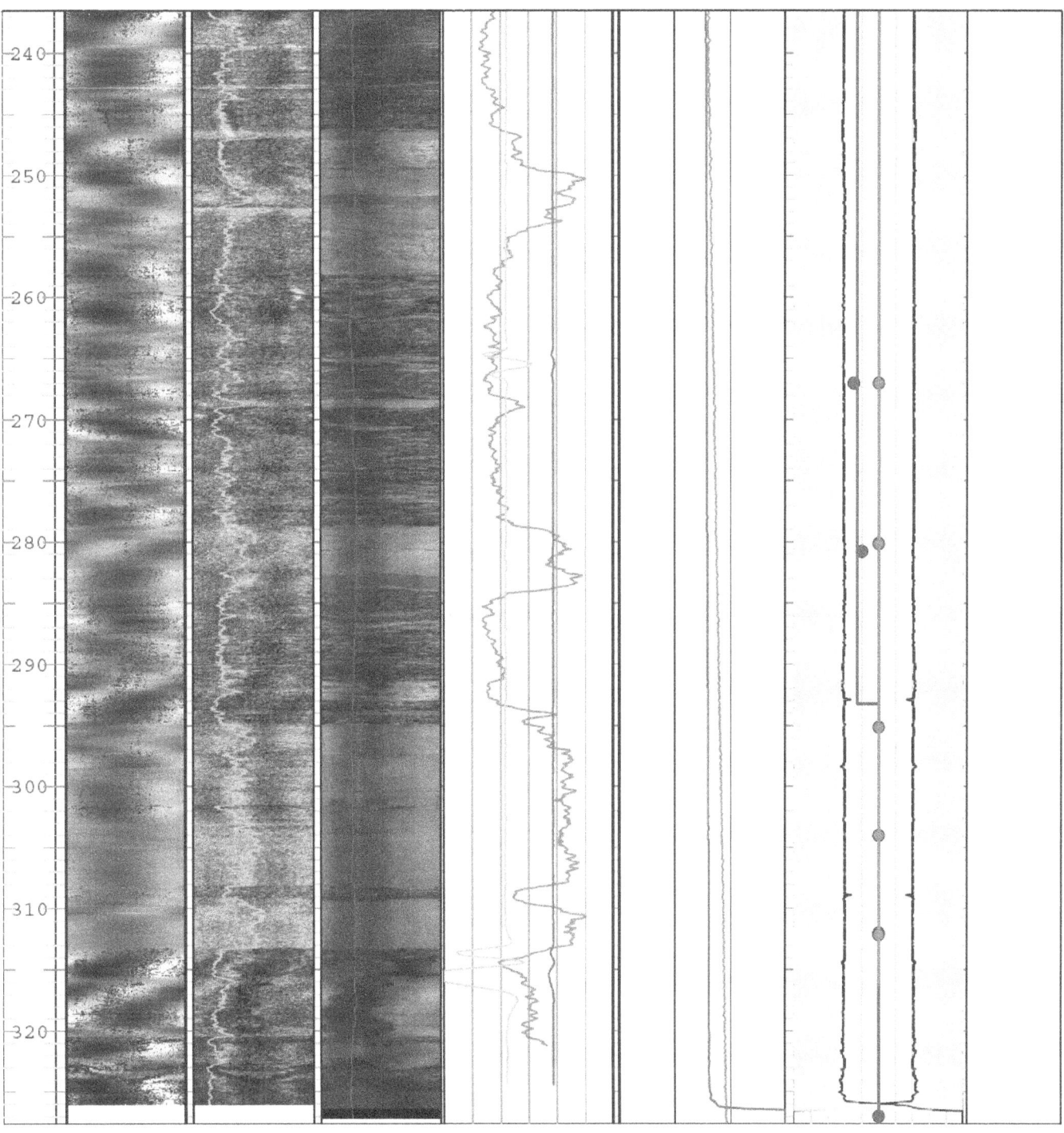

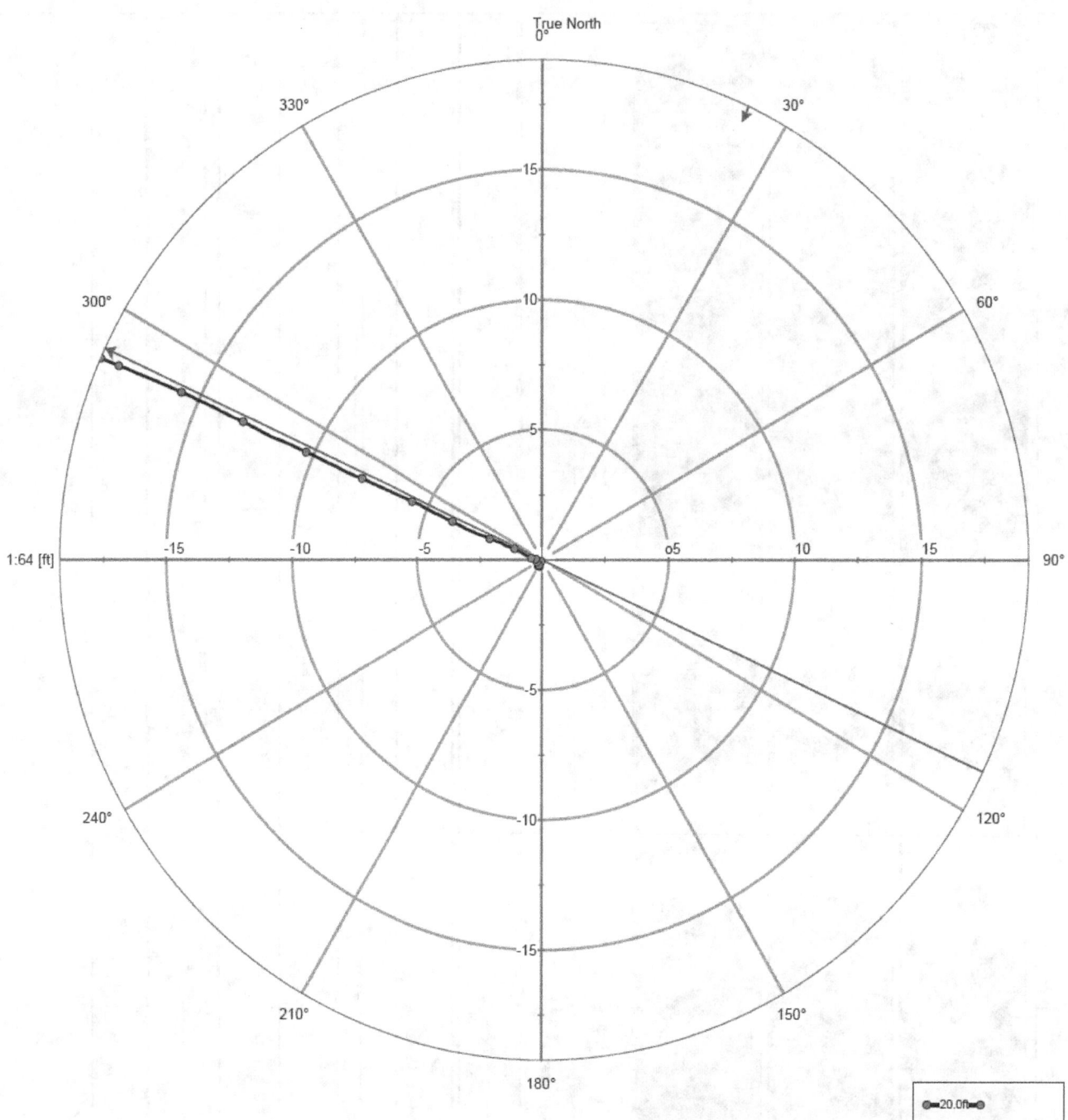

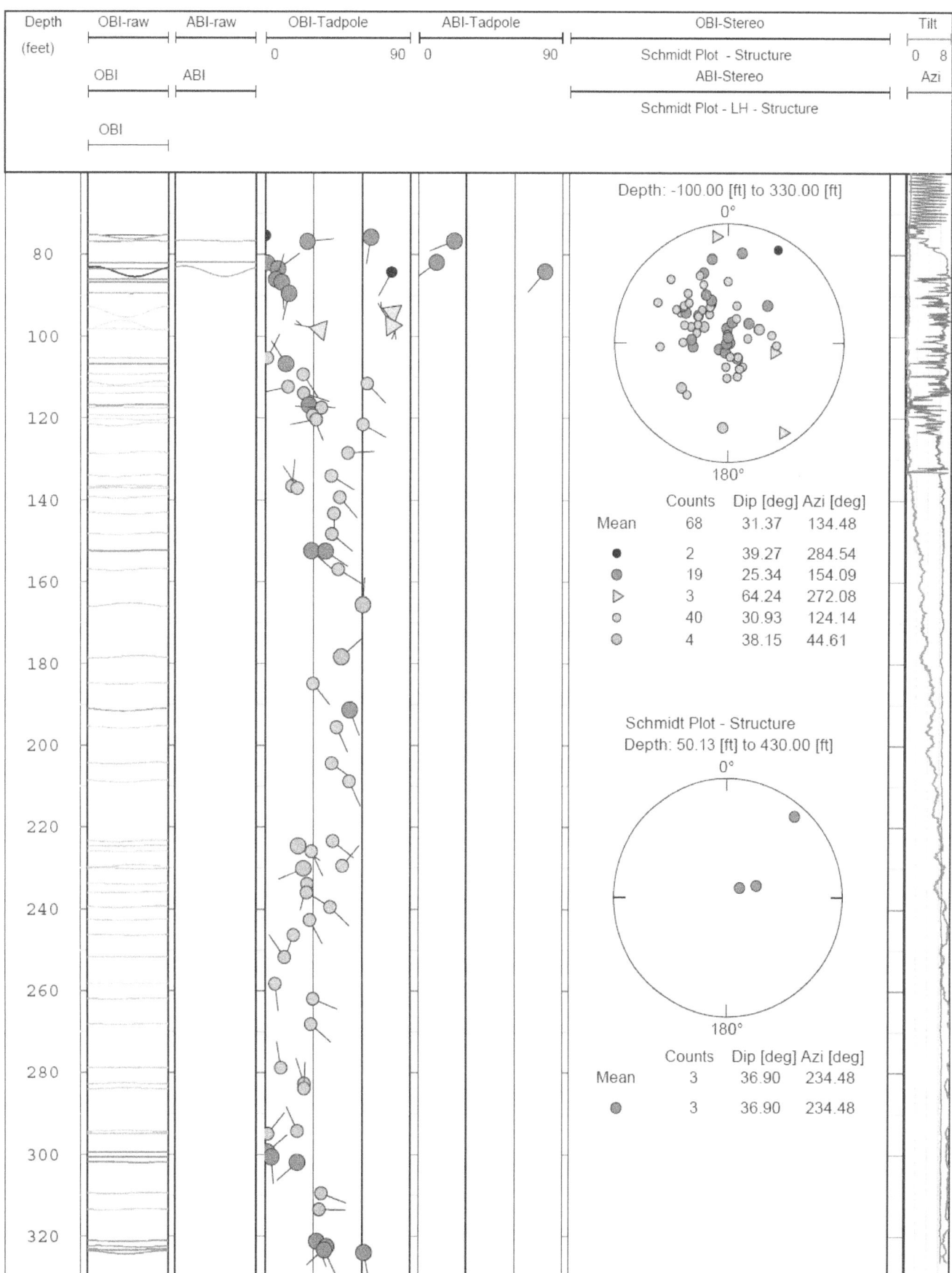

Company USGS   **Site ID** 441226069502201   **Station name** KW-891

Other ID West Gardiner Liberty Lane well   **Date of log**   10/31/2007

County/State Kennebec/Maine   **Office/logging unit** OGW-BG Storrs

Logging operator MJN, CWS   **Observer**   CDJ

Description of log-measuring point (LMP) Top of casing

Height of LMP above/below land-surface datum (in feet) 1.5

Altitude of LMP   169 feet

Borehole depth/diameter/type   347 feet / 6 in. / open

Casing depth/diameter/type   25 feet / 6 in. / steel

Borehole fluid type   Water

Hydrologic conditions   Ambient- Pre-injection conditions

Remarks   Stickup=1.5 foot, Pumping at 0.25-.3 gal/min was conducted as part of stressed flowmeter tests on 10/31/07.

Logs in composite, date and time, manufacturer, serial number (SN), condition

EMI and Gamma, 10/31/07, MSI-2PIA+2PGA, SN: 2307, gamma factory calibrated; count per second used
NE-EMI tool - calibrated after log; data adjusted

Fluid and Caliper, 10/31/07, MSI-2PCA+2FSB, SN: 3009, calibration checked after log; data adjusted to calibration values

OBI-40-Mk4,   10/31/07, MSI-ALT, MSI SN: 073612, Alt SN:061101, tilt and direction check at time of log

ABI-40,   10/31/07, MSI-ALT, MSI SN: 3078, Alt SN: 020906, tilt and direction check at time of log

HPFM,   10/31/07, MSI 2293, SN: 2060, calibration values updated in MsHeat for tool 2060
ambient and pumping logs collected

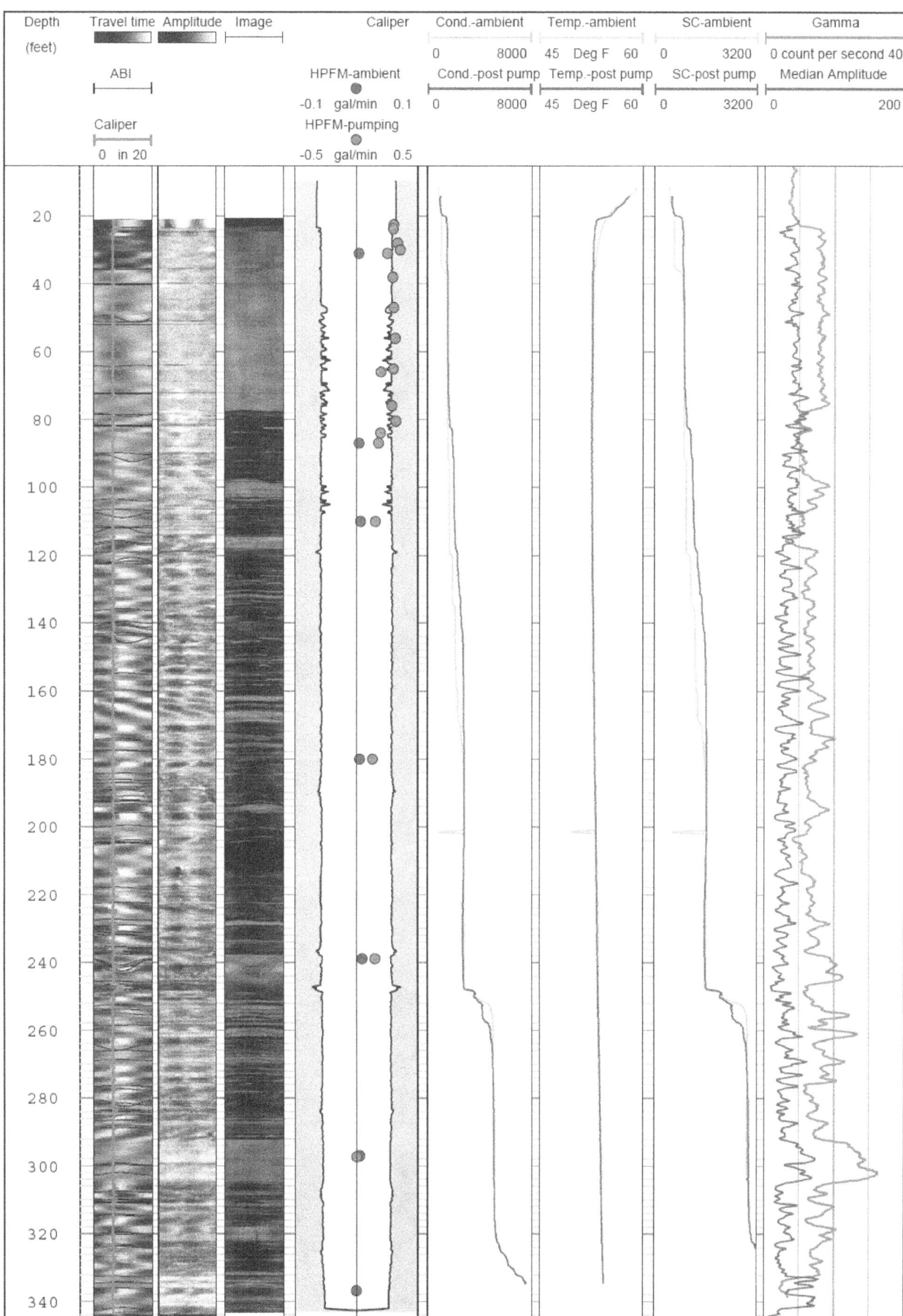

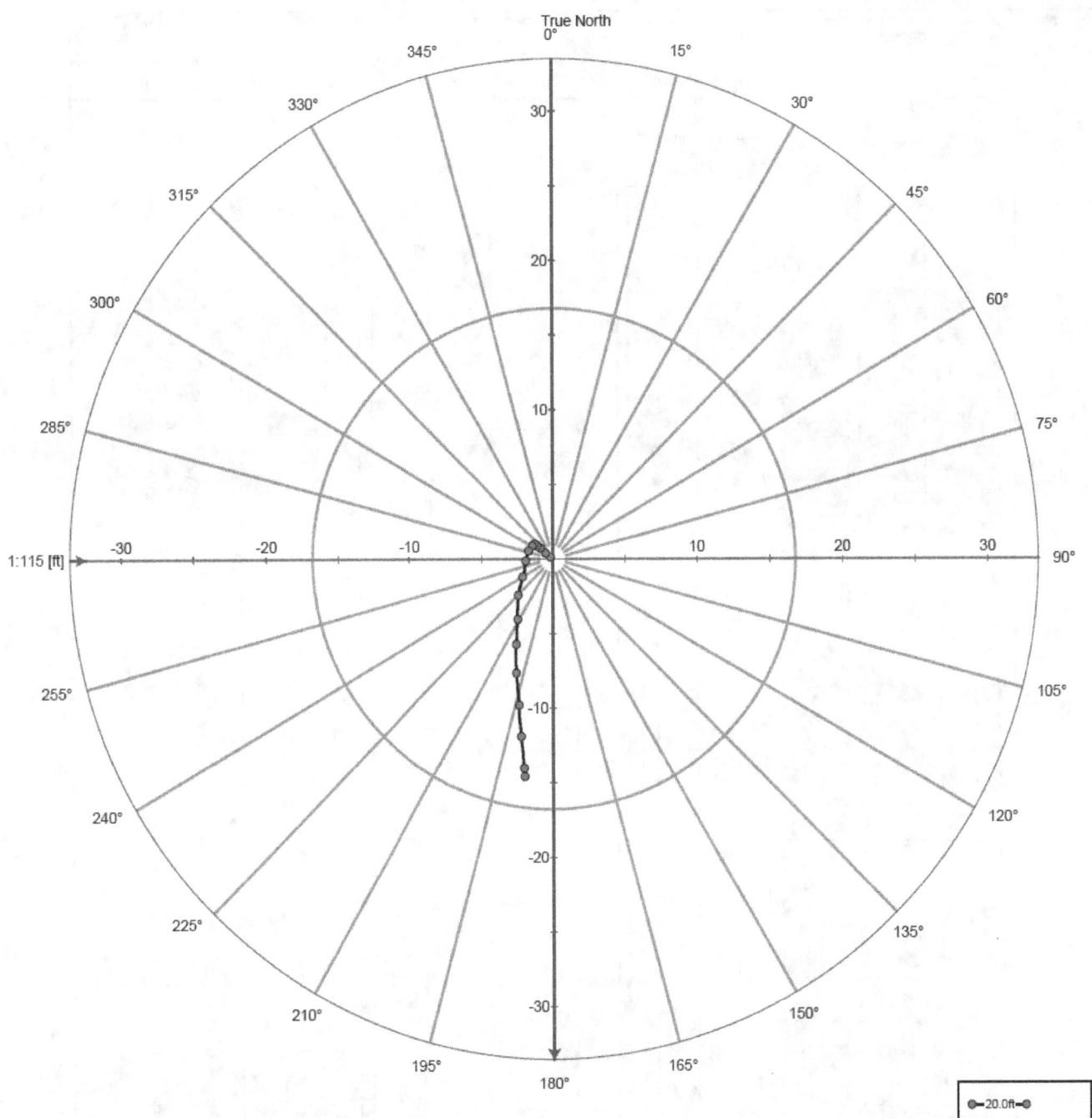

Company USGS   **Site ID** 435103070191701   **Station name**   HW-181

Other ID South Sullivan well   **Date of log** 6/25/08

County/State Hancock /Maine   **Office/logging unit** OGW-BG Storrs

Logging operator MJN, CWS   **Observer**   CDJ

Description of log-measuring point (LMP) Top of 6-inch steel casing

Height of LMP above/below land-surface datum   2.0 feet

Altitude of LMP   46.13 feet   **Magnetic declination**   ~-14 deg

Borehole depth/diameter/type   245 feet / 6 in. / open

Casing depth/diameter/type   40 feet / 6 in. / steel

Borehole fluid type   Water

Borehole fluid depth and date   37.02 feet at 9:54 on 6/25/2008

Hydrologic conditions   Ambient with possible tidal influence

Remarks   120 gallons removed =~80 feet of water column, pulling water from
            205 feet to 125 feet
            Fractures at 43 feet, 186 feet, 198 feet, and 207 feet are hydraulically
            active

Logs in composite, date and time, manufacturer, serial number, condition

EMI and Gamma, 06/25/08, MSI-2PIA+2PGA, SN: 2307, factory calibrated
                    NE-EMI tool - calibrated at time of log

Fluid and Caliper,   06/25/08, MSI-2PCA+2FSB, SN: 3009, calibration checked at time of log

OBI-40-Mk4,   06/25/08, MSI-ALT, MSI-SN 073612 Alt SN:061101, tilt and direction check at time of log

ABI-40,   06/26/08, MSI-ALT, MSI SN:3078, ALT SN: 020906, tilt and direction check at time of log

HPFM,   06/26/08, MSI 2293, SN:2060, calibration values updated in MsHeat for tool 2060
                    ambient and pumping logs collected

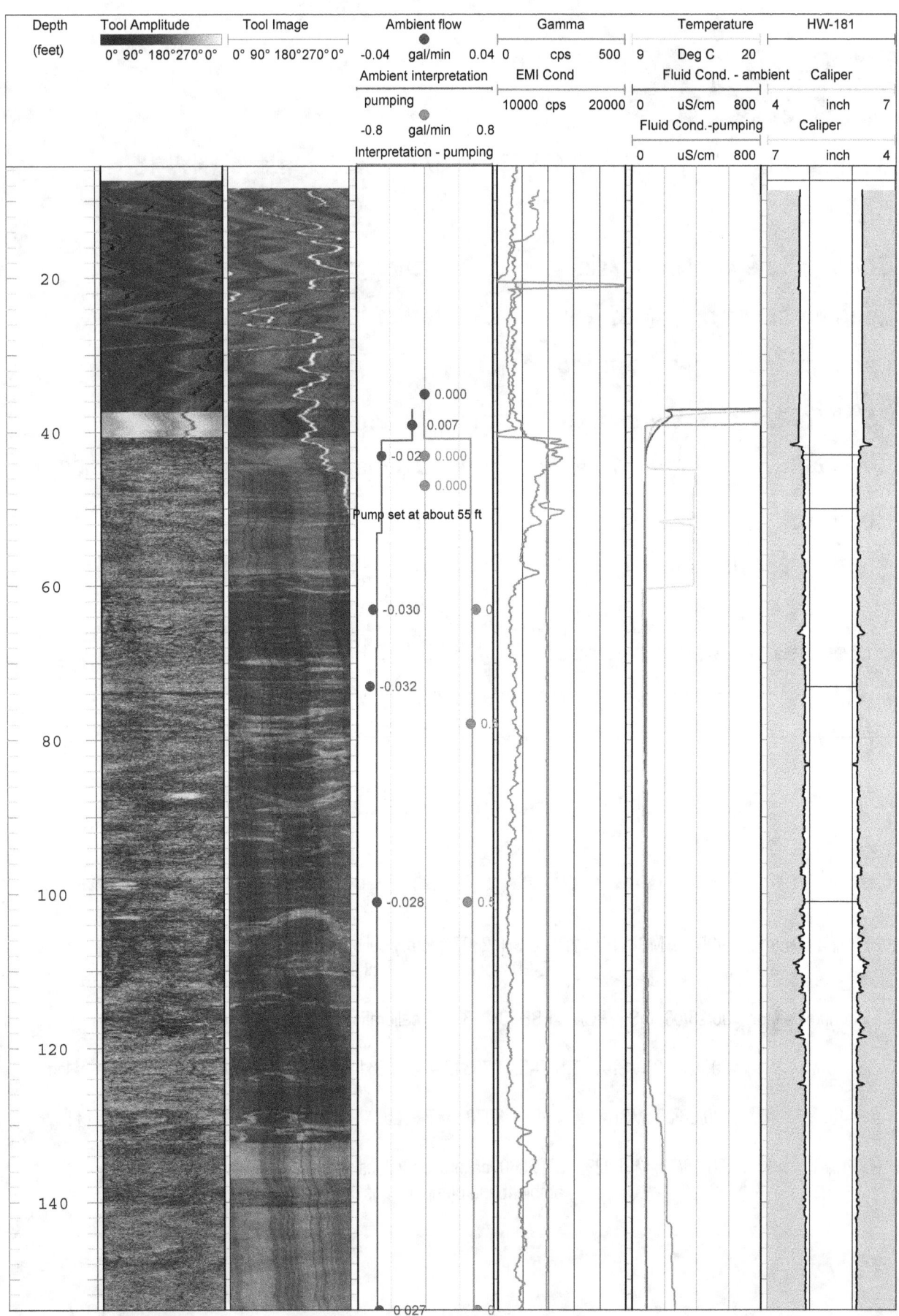

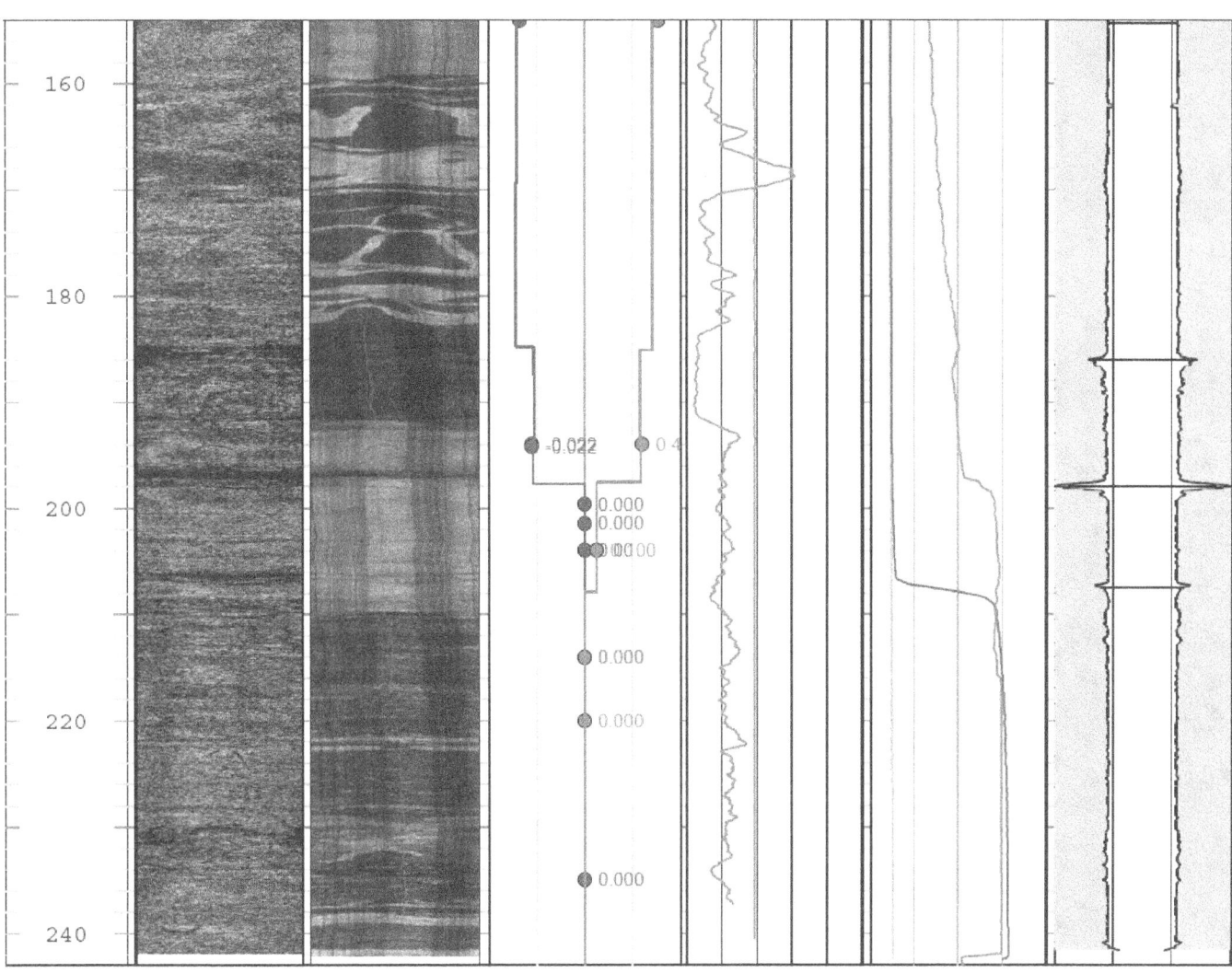

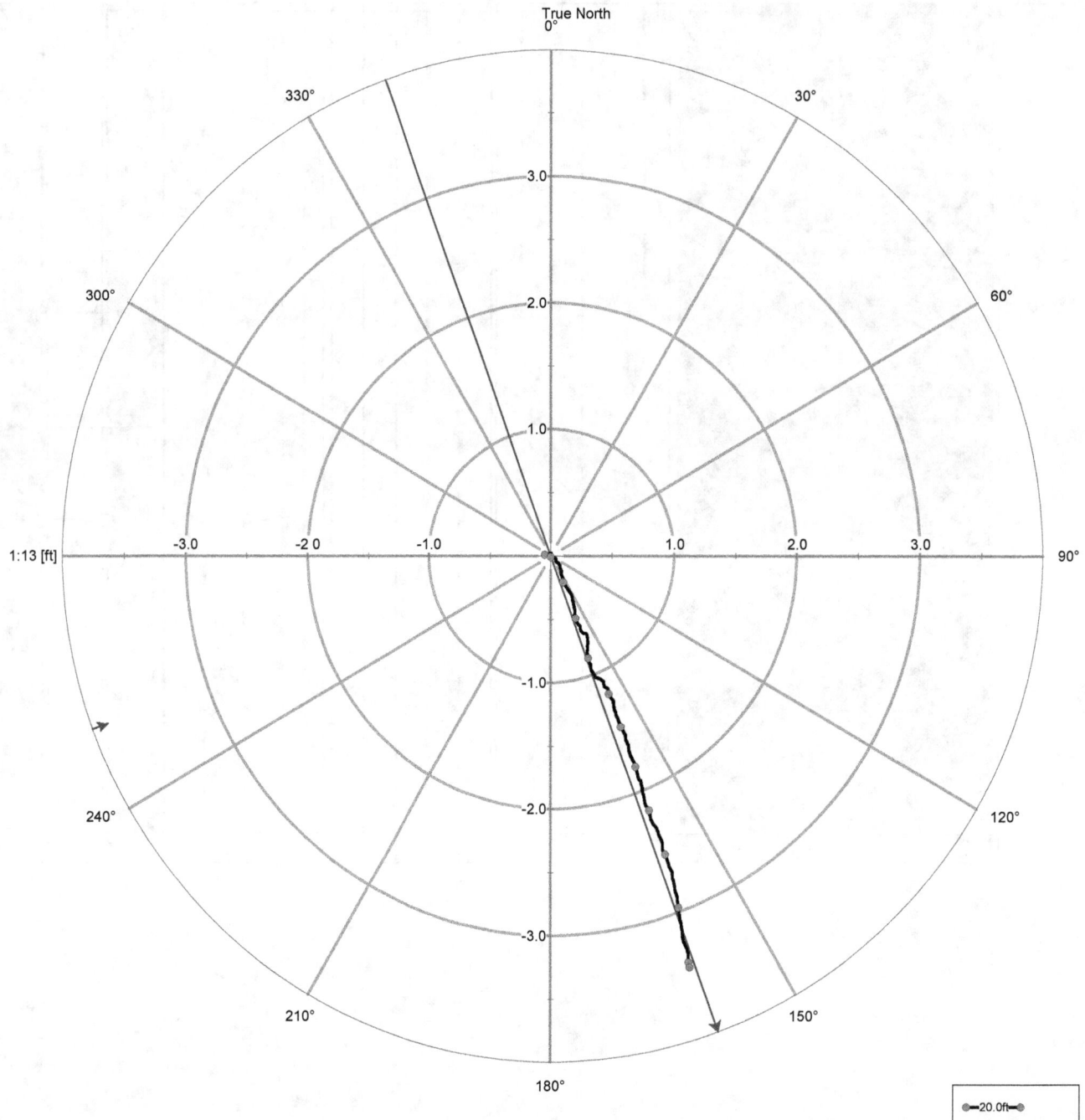

Company USGS   **Site ID** 443145068131801   **Station name** HW-180

Other ID North Sullivan well   **Date of log** 06/27/2008

County/State Hancock /Maine   **Office/logging unit** OGW-BG Storrs

Logging operator C. Schalk   **Observer** C. Johnson

Description of log-measuring point (LMP) Top of casing

Height of LMP above/below   0.5 feet

Altitude of LMP 30.5 feet   **Mag** **declination** ~14

Borehole depth/diameter/type   97.16 feet / 6 in. / open

Casing depth/diameter/type   15.2 feet / 6 in. / steel

Borehole fluid type   Water

Borehole fluid depth and date   15.50 feet at 8:17 on 06-27-2008

Hydrologic conditions Ambient and Pumping

Remarks Domestic submersible pump pulled before logging and re-installed
at end of day

Logs in composite, date and time, manufacturer,   , condition

EMI and Gamma, 06/27/08, MSI-2PIA+2PGA, SN: 2307, factory calibrated
EMI tool - calibrated at time of log

Fluid and Caliper,   06/27/08, MSI-2PCA+2FSB, SN: 3009, calibration checked at time of log - ambient
post-pumping log collected after removing 65 gallons of water

OBI-40-Mk4,   06/27/08, MSI-ALT, MSI-SN 073612 Alt SN:061101, tilt and direction check at time of log

ABI-40,   06/27/08, MSI-ALT, MSI SN:3078, ALT SN: 020906, tilt and direction check at time of log

HPFM,   06/27/08, MSI 2293, SN:2060, calibration values updated in MsHeat for tool 2060
ambient and pumping logs collected

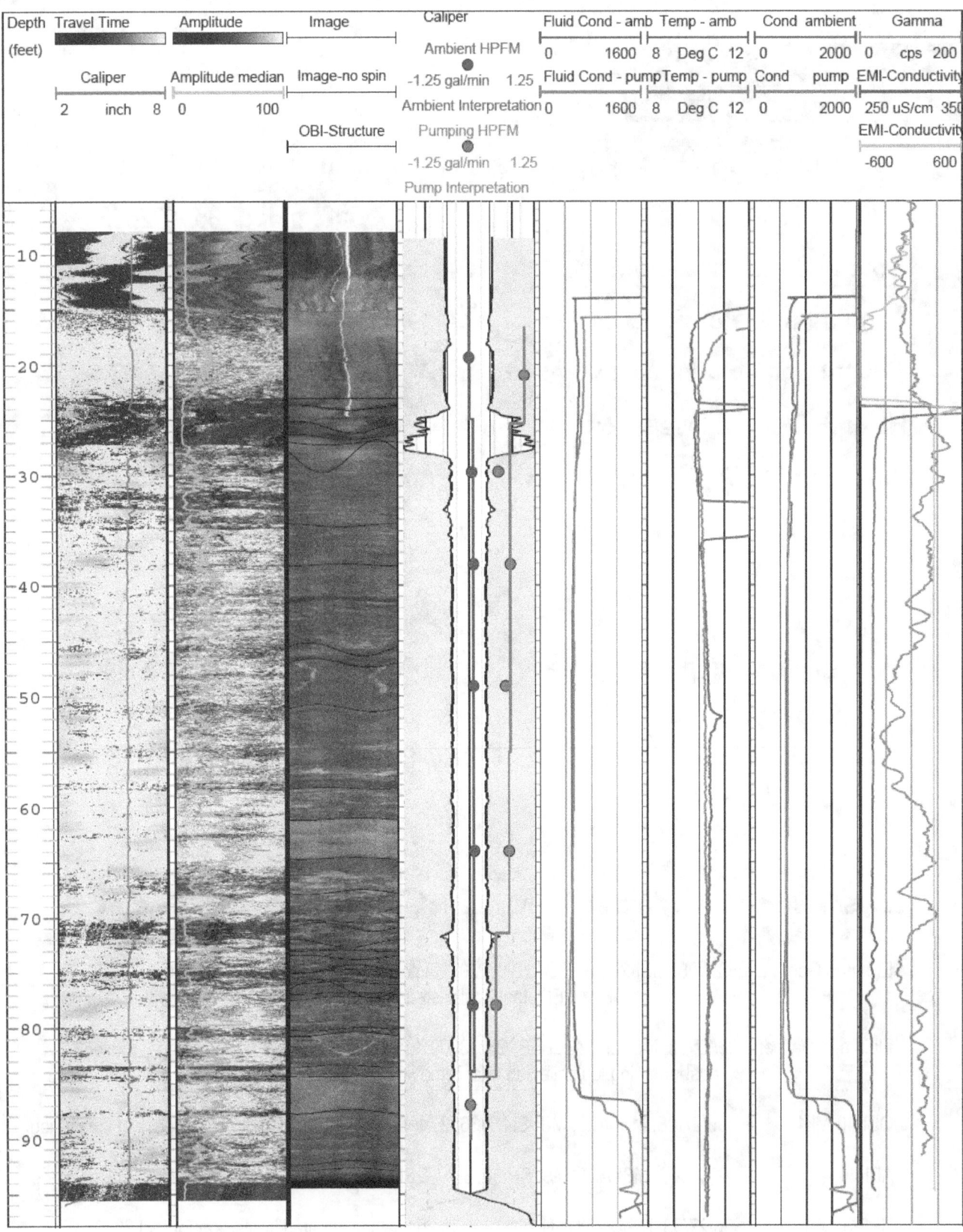

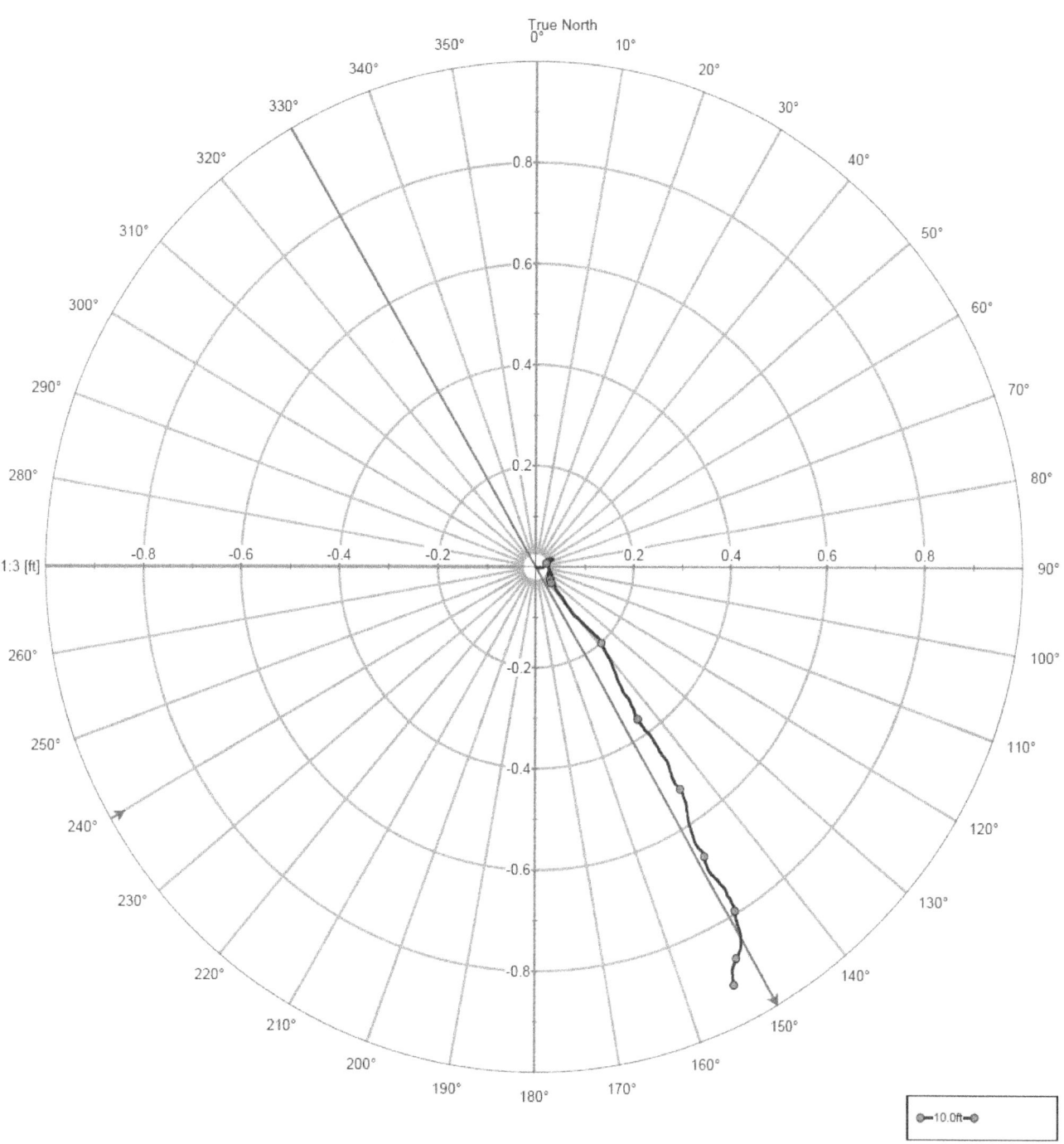

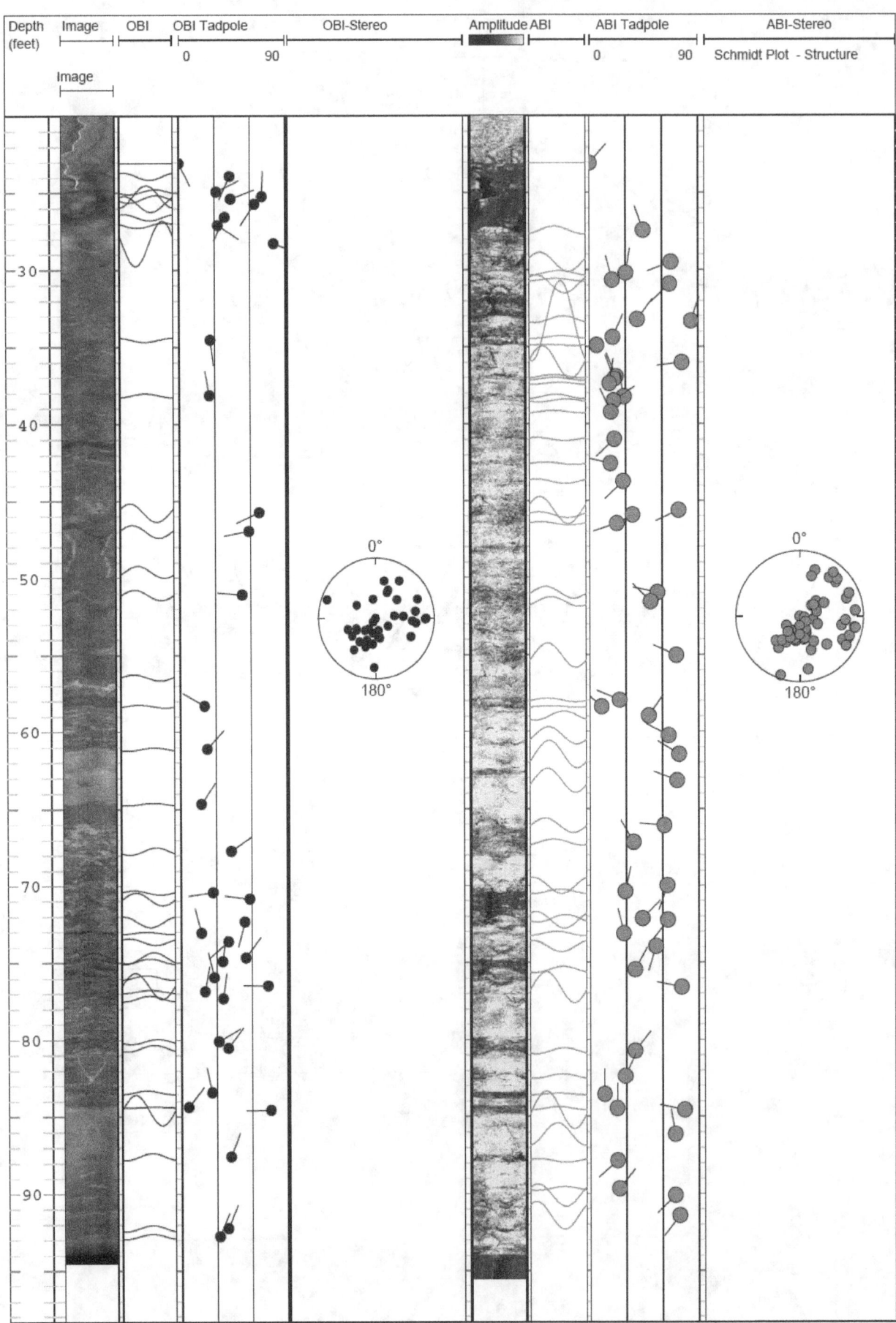

Appendix 2.   Daily Water Levels, Specific Conductance, and Precipitation, All Wells

Figure 2–1.   Graph showing water levels, liquid-equivalent precipitation, and specific conductance in U.S. Geological Survey well 435103070191701 in Gray, Maine, December 2007–May 2009.

Figure 2–2.   Graph showing water levels, liquid-equivalent precipitation, and specific conductance in U.S. Geological Survey well 441226069502201 in West Gardiner, Maine, August 2007–November 2009.

Figure 2–3.   Graph showing water levels, liquid-equivalent precipitation, and specific conductance in U.S. Geological Survey well 443145068131801 in North Sullivan, Maine, September 2007–June 2009.

Figure 2–4.   Graph showing water levels, liquid-equivalent precipitation, and specific conductance in U.S. Geological Survey well 443113068115101 in South Sullivan, Maine, January 2008–June 2009.

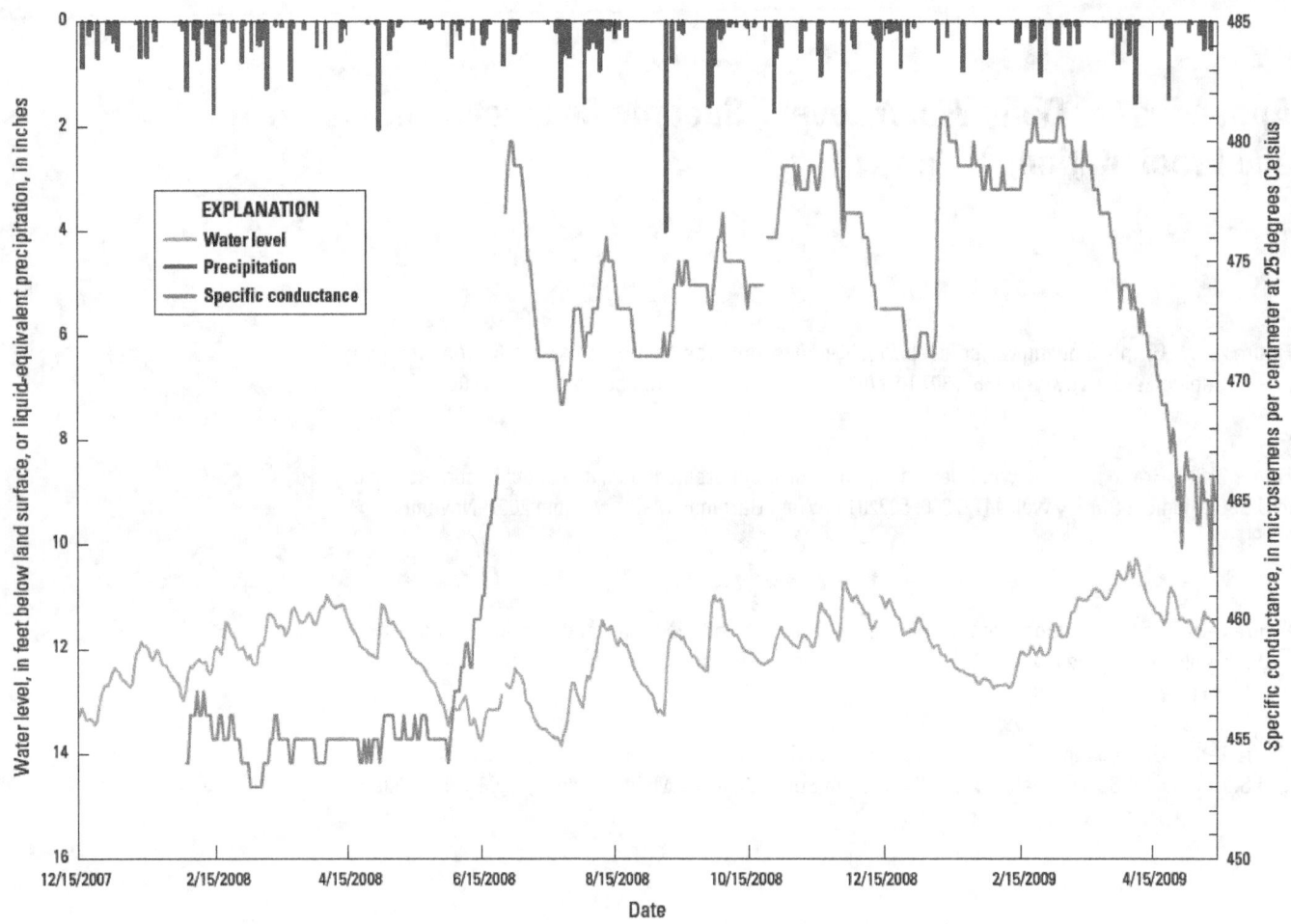

Figure 2–1.   Graph showing water levels, liquid-equivalent precipitation, and specific conductance in U.S. Geological Survey well 435103070191701 in Gray, Maine, December 2007–May 2009.

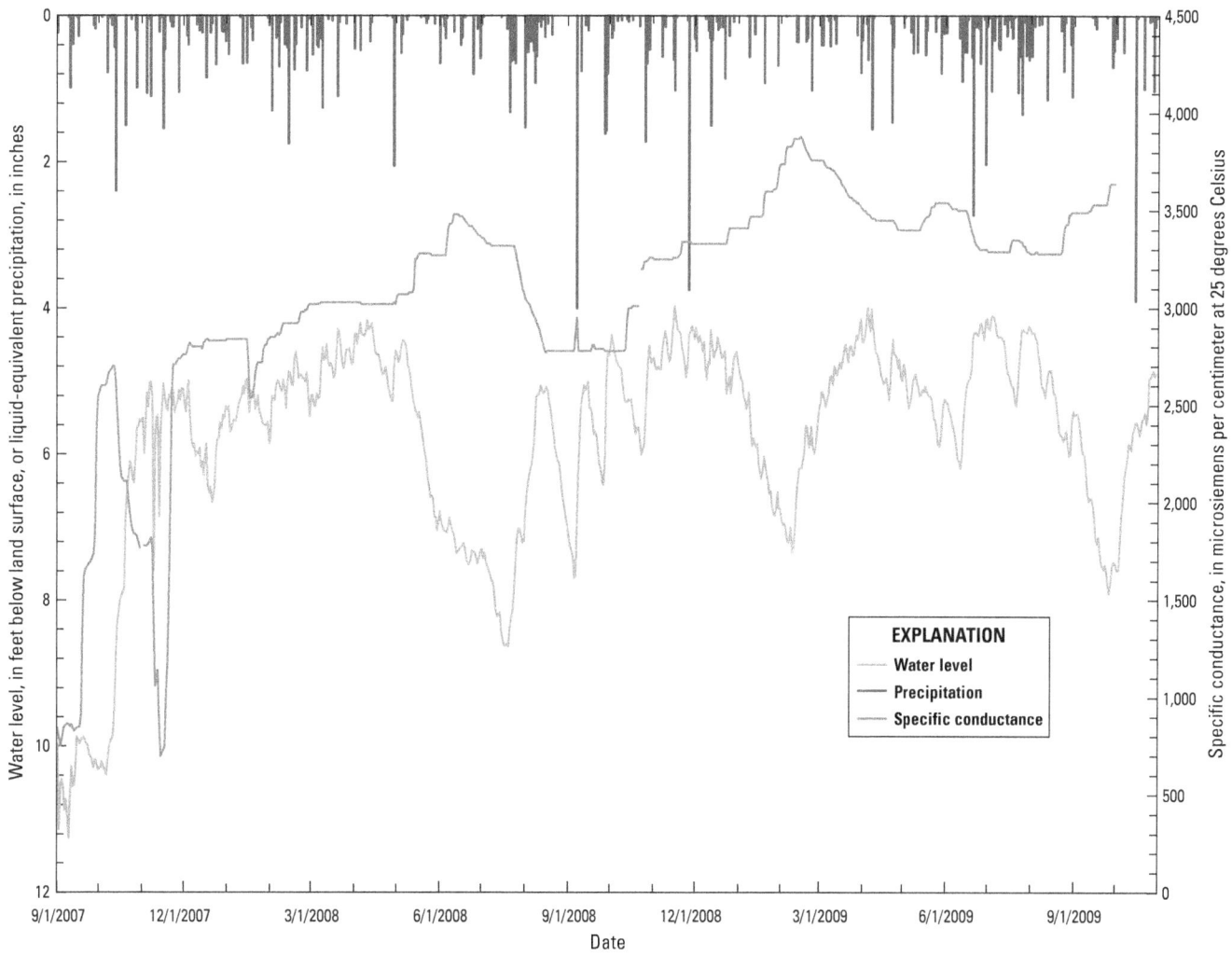

Figure 2–2.   Graph showing water levels, liquid-equivalent precipitation, and specific conductance in U.S. Geological Survey well 441226069502201 in West Gardiner, Maine, August 2007–November 2009.

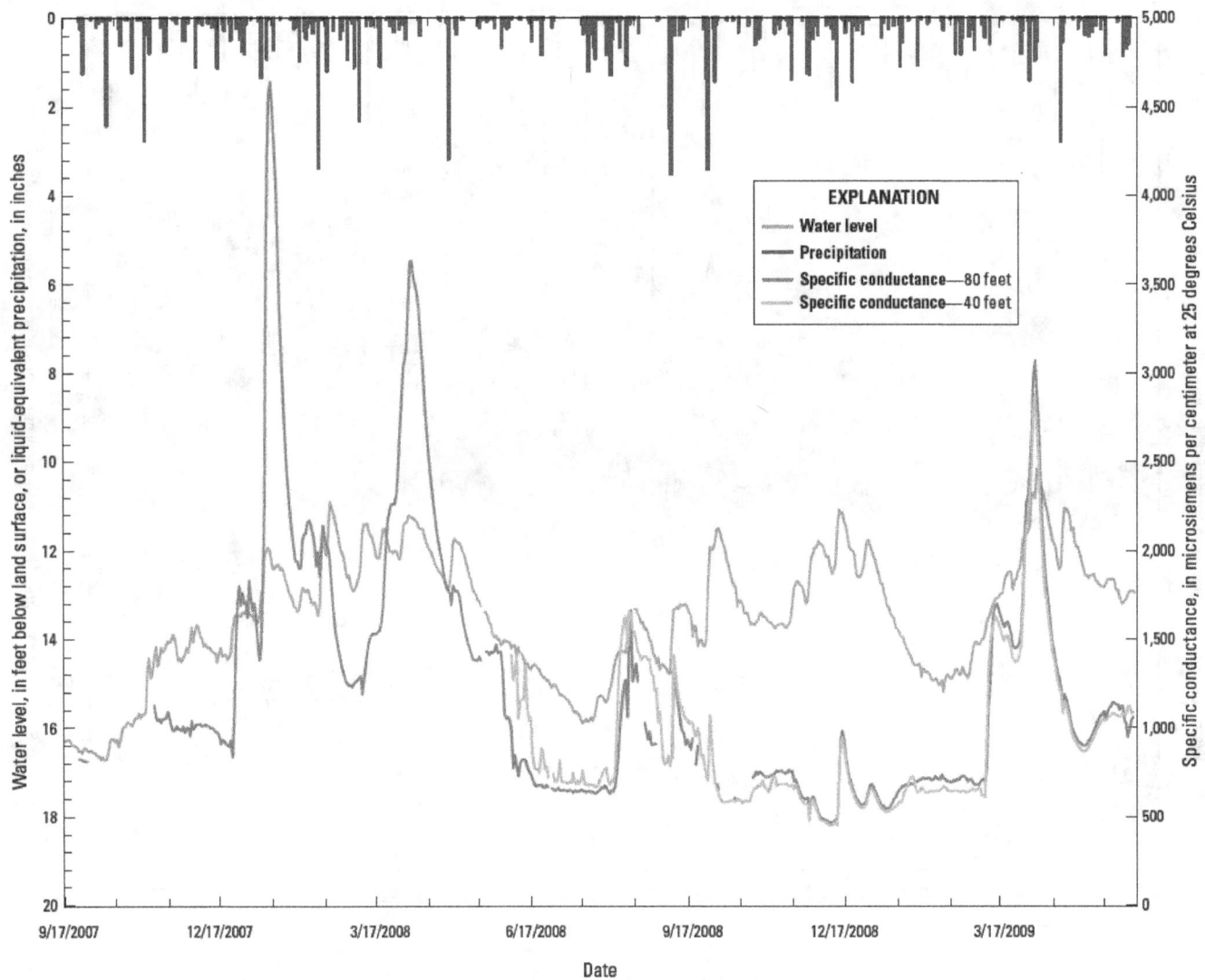

Figure 2–3.   Graph showing water levels, liquid-equivalent precipitation, and specific conductance in U.S. Geological Survey well 443145068131801 in North Sullivan, Maine, September 2007–June 2009.

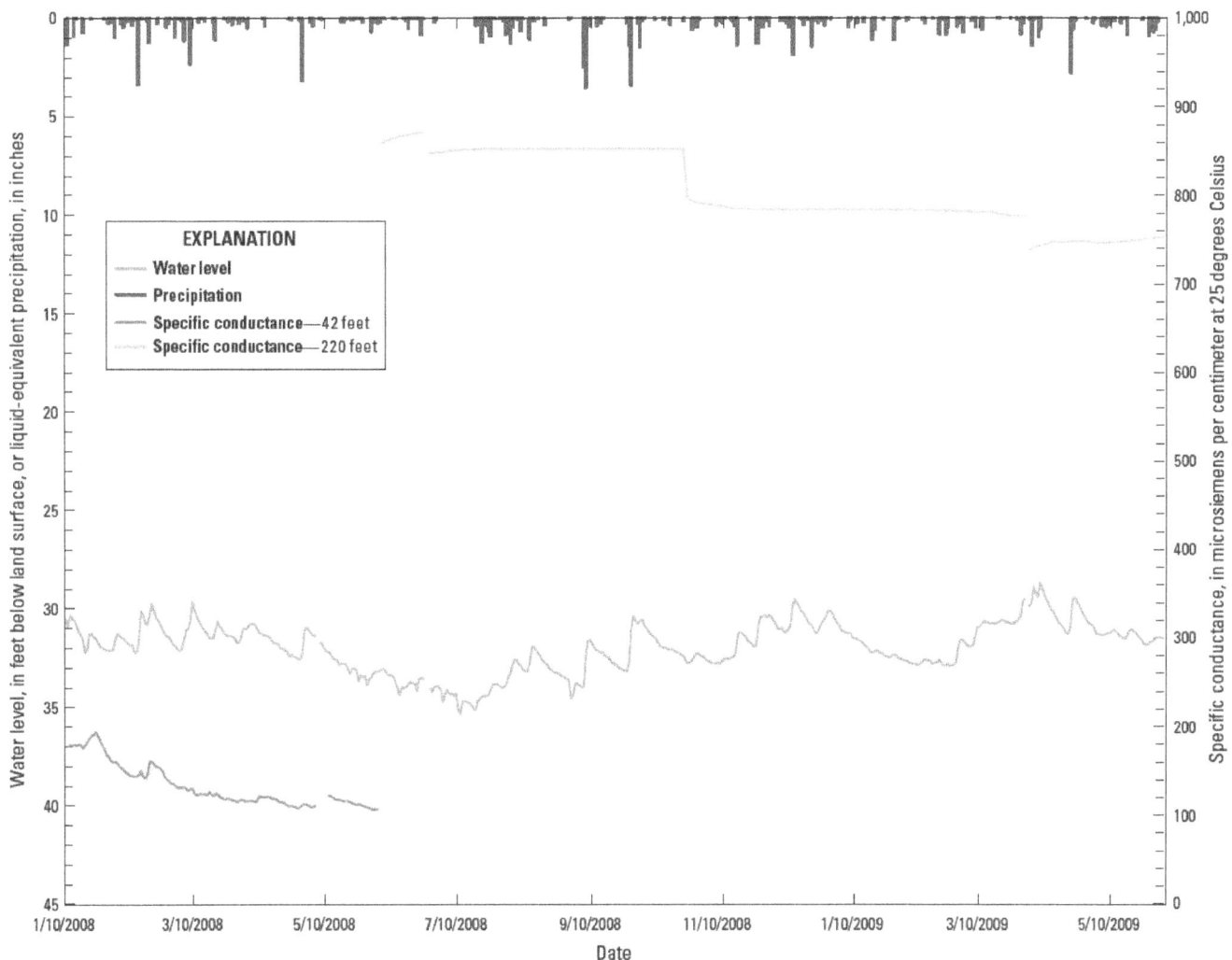

Figure 2–4. Graph showing water levels, liquid-equivalent precipitation, and specific conductance in U.S. Geological Survey well 443113068115101 in South Sullivan, Maine, January 2008–June 2009.

THIS PAGE INTENTIONALLY LEFT BLANK

Prepared by the Pembroke Publishing Service Center.

For more information concerning this report, contact:

Office Chief
U.S. Geological Survey
New England Water Science Center
Maine Water Science Office
196 Whitten Road
Augusta, ME 04330
dc_me@usgs.gov

or visit our Web site at:
http://me.water.usgs.gov

USGS

Schalk and Stasulis—Relations Among Three Variables in Four Road-Salt-Contaminated Wells in Maine, 2007–9—Scientific Investigations Report 2012–5205